SpringerBriefs in Applied Sciences and Technology

T0171843

More information about this series at http://www.springer.com/series/8884

Navin G. Ashar

Advances in Sulphonation Techniques

Liquid Sulphur Dioxide as a Solvent of Sulphur Trioxide

 Springer

Navin G. Ashar
Navdeep Enviro And Technical Services
    Private Limited
Ambernath
India

ISSN 2191-530X   ISSN 2191-5318   (electronic)
SpringerBriefs in Applied Sciences and Technology
ISBN 978-3-319-22640-8   ISBN 978-3-319-22641-5   (eBook)
DOI 10.1007/978-3-319-22641-5

Library of Congress Control Number: 2015947109

Springer Cham Heidelberg New York Dordrecht London

Springer International Publishing AG Switzerland is part of Springer Science+Business Media
(www.springer.com)

During my undergraduate studies at Banaras Hindu University (1950–54), the class was asked to design a "50 TPD Sulphuric Acid Plant".

After postgraduate study in Chemical Engineering at MIT (Cambridge, USA) 1958, I was appointed on the MIT faculty until 1961. Following that I returned to India and joined a fertilizer company 80 km away from Mumbai, namely M/s. Dharamsi Morarji Chemical Co. Ltd. The company had constructed a 50 TPD Sulphuric Acid plant just a year earlier. My attachment to Sulphuric Acid and Sulphur-based chemicals is still intense after over 53 years, to the extent that my wife complains that even today my first love is Sulphuric Acid!

I married Rajni in 1962. Over the five decades she has supported and inspired me to be creative and innovative. This book is the result of her devotion and support.

I dedicate this pathbreaking document to my beloved wife.

Preface

It is about time that a complete in-depth analysis is provided for future researchers and technocrats on the impact of liquid sulphur dioxide and liquid sulphur trioxide to carry out complex and difficult sulphonations.

Most of the sulphonation proceses are exothermic. Liquid sulphur dioxide can act as auto-refrigerant. It can be recycled after condensing. The property of sulphur dioxide indicates that condensing can be carried out at ambient temperatures at pressures of 6–8 kg/cm^2.

The current production techniques can be replaced by the innovative process of using liquid sulphur, liquid sulphur dioxide and liquid sulphur trioxide.

The costly and cumbersome process plant can be simplified with economic advantage and better conversion efficiency.

This book gives a new pathbreaking direction to the production of sulphuric acid as well as sulphonation of organic and inorganic chemicals.

Contents

1 Introduction 1
  1.1 Preamble 1
    1.1.1 History of Manufacture of Sulphuric Acid in India . . . 1
    1.1.2 History of Manufacture of Sulphuric Acid 2
    1.1.3 Salient Features of the Modified (3 + 2)
           DCDA Process 3

2 Chemical and Physical Properties of Sulphur Dioxide
   and Sulphur Trioxide 9
  2.1 Introduction 9
  2.2 Sulphur Dioxide Physical Properties 10
  2.3 Vaporisation of SO_2 10
  2.4 The Solubility of SO_2 in Sulphuric Acid 11
  2.5 Solubility of Sulphur Dioxide in Water 12
  2.6 Chemical Properties of Sulphur Dioxide 12
  2.7 Physical Properties of Sulphur Trioxide 13
  2.8 General Properties of Liquid Sulphur Trioxide 13
  2.9 Properties of Liquid Sulphur Trioxide 14
  2.10 Viscosity of Liquid Sulphur Trioxide 15
  2.11 Specific Gravity of Sulphur Trioxide 15
  2.12 Vapour Pressure of Liquid Sulphur Trioxide 15
  2.13 Molar Heat Capacity of Liquid Sulphur Trioxide 16
  2.14 Vaporisation Curves for Sulphur Dioxide 16
  2.15 Enthalpy of Sulphur Trioxide Gas 16
  2.16 Chemical Properties of Sulphur Trioxide 17
    2.16.1 Commercially Sulphur Trioxide Is Produced
            by Converting 10–12 % SO_2 by Catalytic
            Conversion at Temperatures Between
            360–600 °C in Multipass Converter of Sulphuric
            Acid Plant 17

2.17   One of the Special Chemical Properties of SO_3
        Which Has Been Safer but not Explored till date   18
2.18   Sulphur Trioxide Is a Strong Sulphonating Agent
        for Difficult, Organic and Inorganic Chemicals.   19
        2.18.1   Treatment of Sulphuric Acid Plant Tail Gas
                 for Final Absorption Tower   19

3   Manufacture of Sulphonating Agents Such as 25 and 65 %
     Oleums as well as Liquid Sulphur Trioxide   21
     3.1   Introduction   21
     3.2   Production of 25 % Oleum   21
            3.2.1   History   21
     3.3   Technical Considerations   22
     3.4   Manufacturing   23
     3.5   65 % Oleum   23
            3.5.1   Introduction   23
     3.6   Manufacturing   24
     3.7   Uses   24
     3.8   Sulphur Trioxide (Liquid or Gas)   24
            3.8.1   Introduction   24
     3.9   Manufacture.   25
     3.10   Economic Considerations.   25

4   Manufacture of Liquid Sulphur Dioxide   27
     4.1   Manufacture of Liquid Sulphur Dioxide.   27
     4.2   Thermodynamic and Kinetic Consideration
            of the NEAT's Process   28
     4.3   International Scenario   29
     4.4   Merchant Market for SO_2 in Various for Many Industrial
            Applications.   31
     4.5   Process Description.   31
     4.6   Operational Considerations.   32
            4.6.1   Condensation and Filling Section   32
     4.7   Economics.   32
     4.8   Environmental Considerations.   33
     4.9   Conclusion.   33

5   World Production of Liquid SO_2 and SO_3   35
     5.1   Introduction   35
     5.2   World Scenario.   35
            5.2.1   Comparative Analysis on Techno Economic
                     Considerations   35

5.3   Economics of Manufacture of Liquid SO_2   38
5.4   Economics of Manufacture of Liquid SO_3   38
5.5   Conclusion.   40

6   **Techno Economic Evaluation of Processes Involved
     to Manufacture Liquid Sulphur Dioxide and Liquid
     Sulphur Trioxide**   41
     6.1   Introduction   41
     6.2   History   42
     6.3   Production by Burning Sulphur Cooling, Absorption
            in Alkali and Desorption, Drying by Sulphuric Acid,
            Compression and Condensation by Refrigeration.   42
     6.4   Production by Use of Organic Solvent from by Product
            SO_2 Generated in Specific Chemical Reactions.   43
     6.5   Production by Use of Concentrated Oleum (65 %)
            and Solid Sulphur Using Compression and Refrigeration
            (Batch Process).   43
     6.6   Production by Using Molten Sulphur and Liquid SO_3
            Under Pressure Without Compression and Refrigeration
            (Adopted by NEAT)   43
     6.7   Economic Considerations.   44
     6.8   Conclusion.   44

7   **Application of Sulphonation by Liquid SO_3 Dissolved
     in Liquid SO_2**.   45
     7.1   Introduction   45
     7.2   Properties of Sulphamic Acid.   45
     7.3   Process   46
     7.4   Uses   50
     7.5   Conclusion.   50

8   **Impact on the Future Processes for the Manufacture
     of Chemicals**.   51
     8.1   Introduction   51
     8.2   Raw Materials Required   51
     8.3   Major Areas in Which Experimental Work Should
            Be Directed   52
     8.4   Specifications (PTSA Monohydrate)   52
     8.5   Commercial Details Current Manufacturers in India   53
     8.6   Applications and End Use   53
     8.7   Effluent Expected   53
     8.8   Alternate Process (Proposed)   54
     8.9   Key Physical Properties for the New Process   54

9   Case Studies and Its Commercial Application 55
    9.1   Introduction 55
    9.2   NEAT's Innovative "Cold Process". 56
    9.3   Para Toluene Sulphonic Acid 56
    9.4   Synthesis (Outline) 57
    9.5   Specifications (PTSA Monohydrate) 57
    9.6   Commercial Details. 57
    9.7   Some Key Physical Parameters. 58
    9.8   Important Inferences 58
    9.9   Techno Commercial Advantages of NEAT's Innovative
          Process to Manufacture PTS Acid. 59
    9.10 Case Study for Innovative Process Invented by NEAT
          to Carry Out Chloro Sulphonation of Toluene
          Without Refrigeration and CSA Plant as Raw Material
          to Manufacture Saccharine 59
          9.10.1 Introduction 59
    9.11 Conventional Process 59
    9.12 A Brief Description of the Innovative Process
          Is to Affect the Drawbacks in the Conventional Process
          Is Indicated in Fig. 9.1. the Main Features of the Process
          Are as Under 60
    9.13 Economics 60
    9.14 Conclusion. 62

10 Summary 69

Appendices 71

Bibliography 91

Chapter 1
Introduction

1.1 Preamble

Science and technology can never remain stagnant. Innovation and research has been the driving force for progress. Sulphur based chemicals particularly Sulphuric Acid manufacturing process is no exception. Over the past two centuries the technology to produce Sulphuric acid and sulphonating agents has witnessed many milestones which are outlined in detail in this chapter.

In the 21st century an attempt is made in this presentation of a new, unconventional and innovative approach to produce sulphuric acid and sulphonating agent without a sulphur furnace, multipass converters, heat exchangers or expensive towers. A special feature of this process will be zero emission of sulphur dioxide, eliminating more than one million tonnes of acid rain at the current production of Sulphuric Acid in the world.

This process will also provide a new direction to sulphonation processes using liquid sulphur dioxide and liquid sulphur trioxide.

1.1.1 History of Manufacture of Sulphuric Acid in India

Sulphuric acid in India became the key for the manufacture of chemicals in the twentieth century. In the year 1919, there were three Sulphuric acid plants imported from Monsanto (USA) including nuts and bolts and gaskets, each having capacity of only 10 tonnes/day (TPD).

The plants were imported by M/s. Dharamsi Morarji Chemical Co. (DMCC) Ltd at Ambernath, M/s. Punjab Chemicals Ltd and M/s. DCM at Delhi.

© The Author(s) 2016
N.G. Ashar, *Advances in Sulphonation Techniques*,
SpringerBriefs in Applied Sciences and Technology,
DOI 10.1007/978-3-319-22641-5_1

Subsequently three more plants were added of similar capacity by M/s. Eastern Chemicals at Chembur, M/s. Perry at Ranipeth (Tamilnadu) and M/s. Bengal Chemicals Ltd at Calcutta.

Due to discouraging policies of the British government, the above plants could not expand since the import of chemicals using Sulphuric Acid from UK was cheaper than chemicals manufactured in India.

It was only after independence in 1947 that the expansion of the manufacturing capacities of sulphuric acid was necessary and encouraged by the Govt. of India.

1.1.2 History of Manufacture of Sulphuric Acid

1.1.2.1 Sulphuric Acid Manufacture has Flourished Since the Mid-19th Century

Sulphuric acid industry got a head start in the 1940s due to the invention of Vanadium Pentoxide as catalyst to convert Sulphur dioxide to Sulphur trioxide popularly known as the "Contact Process". This enabled industry to put up large Sulphuric acid plants of higher concentration than achieved by prevalent lead chambers. The high concentration was required to produce phosphoric acid and phosphatic fertilisers.

In 1960s the environment protection laws became stringent and modified the single contact single absorption (SCSA) plants. The conversion efficiency of SCSA process giving 96–96.5 % conversion produced SO_2 emission of 16–20 kg/tonne of Sulphuric acid produced. This would result in high quantity of acid rain affecting the environment. Thus it became necessary to innovate the manufacturing process by the use of double catalyst double absorption (DCDA) giving 99.5 % efficiency conversion. This DCDA system became popular for single plants with capacity in excess of 1000 TPD.

The rise in energy costs due to the formation of OPEC in mid 1970s raising the crude oil price from USD 8 to USD 60 per barrel required further innovation. Cogeneration was introduced using byproduct steam to produce electricity. In the 70s and 80s, many sulphuric acid plants incorporated cogeneration with economic advantage. In view of this, the minimum capacity that is economically attractive has become 500 tonnes per day and above. For example, the exothermic energy available per 100 tonne of acid can give us a maximum of 1.8 MW of electricity with the 'WHRS' system developed by Monsanto Chemicals of USA. A 500 TPD Sulphuric acid plant with DCDA process can easily provide 5–6 MW cogenerated electric power.

1.1.3 Salient Features of the Modified (3 + 2) DCDA Process

In order to reduce the emission of sulphur dioxide to below ½ kg/tonne of 98 % Sulphuric acid, the conventional DCDA (3 + 1) system is modified as follows: (Please see Fig. 1.1).

In Fig. 1.2 the sulphur dioxide emission as a function of SO_2 concentration at converter inlet and SO_2 conversion rate are shown.

1. First Pass (as upper ignition layer) and Fifth pass (entirely) of the catalyst bed will have Cesium promoted catalyst. This is to ensure faster pick-up of conversion process after any stoppage and also to achieve closer approach to equilibrium conversion of SO_2 to SO_3. Instead of the conventional (3 + 1) DCDA Process, (3 + 2) DCDA Process shall be used. In this innovation, after all the SO_3 produced in the first three passes has been absorbed in the Oleum Towers and the IPAT, the gases shall be led through two more catalyst passes instead of the usual one pass only. A dry air injection facility shall be provided at the outlet of the fourth pass to increase the oxygen content of the gases which will help in shifting the equilibrium of the reversible reaction:

$$2\,SO_2 + O_2 \rightarrow 2\,SO_3 \Delta H = -74.3 \, Kcal/g\,Mole$$

   The fifth pass shall contain Cesium promoted catalyst which has a lower ignition temperature of 360 °C. As lower temperature favours higher overall conversion, it is obvious that a higher yield of the above reaction will be obtained as compared to the conventional (3 + 1) DCDA Process.

   The first pass of the converter will have as the upper half (approximately) Cesium based catalyst which has a low ignition point of 360 °C as compared to 410.0–420.0 °C for the conventional catalyst. This will allow the cooling of the WHB#1 outlet gases by a further 50.0–60.0 °C and thus producing some more steam. Another advantage is the faster pickup of conversion after any Plant stoppage.

2. A higher (10.0–10.5 %) SO_2 gas strength in the burner outlet gases will be possible due to use of Cesium promoted catalyst. This will also require lower volumes of gases to be handled, thus reducing power consumption.

3. Cesium activated catalyst is used to give stack emissions of 150 PPM SO_2 as against 500–700 PPM permitted by Environment Protection Agency of USA.

4. Air injections shall be provided at appropriate places in the conversion system.

5. A separate acid circuit shall be provided for Final Absorption Tower to minimise SO_2 content in the stack (Please see Fig. 1.3).

6. Drying Tower and Interpass Absorption Tower will have a common acid circuit.

7. Separate Plate Heat Exchangers shall be provided for FAT and (DT + IPAT).

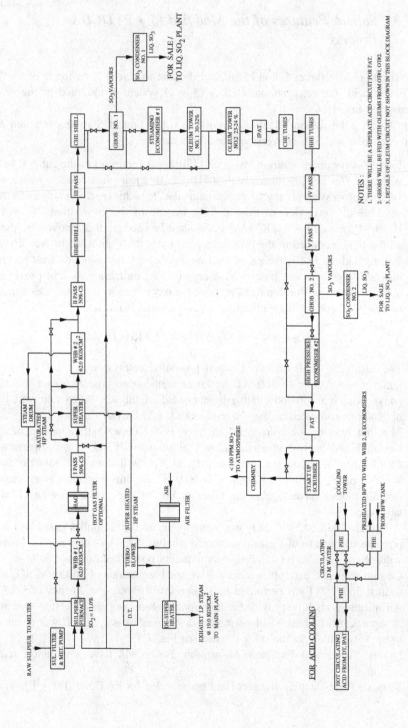

Fig. 1.1 Block diagram for sulphuric acid/oleum/SO₃ plant (3 + 2 DCDA process) with cogeneration

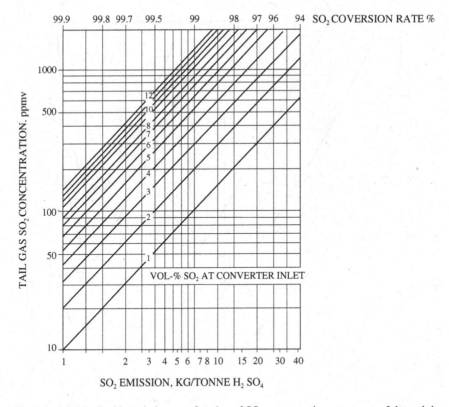

Fig. 1.2 Sulphur dioxide emission as a function of SO$_2$ concentration at converter Inlet and the SO$_2$ conversion rate

8. A Sulphur Filter along with a standby filter can be also provided to ensure that only filtered sulphur is fed to the plant. This has the following advantages:

   • Fouling of heat transfer surfaces of boilers and heat exchangers is minimised
   • Wear-and-tear of sulphur pumps is reduced
   • Active surfaces of catalysts do not get masked by dust, and hence, the conversion of SO$_2$ to SO$_3$ does not deteriorate with time.
   • Steady process conditions can be maintained due to which product acid and oleum quality is better.
   • Lastly, rate of build-up of pressure drop over time is reduced. Hence, power consumption does not increase.

9. Two Centrifugal Pumps are generally provided for filtering the raw sulphur. Metering Pumps are provided to ensure steady Sulphur feed to the burner. This will help in maintaining steady process conditions very essential for efficient Plant operations.

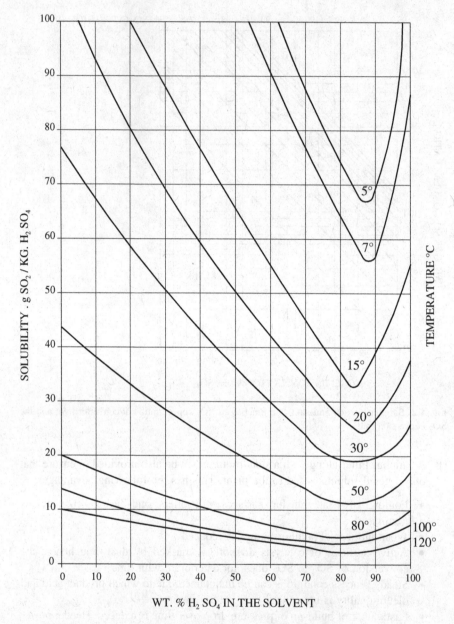

Fig. 1.3 Solubility of sulphur dioxide in sulphuric acid at p (SO₂) 1,013 mbar and at various temperature

10. The Sulphur burner shall be lined with insulating bricks, high alumina fire bricks and very high alumina (60–65 %) fire bricks to conserve heat inside and to enable operation of the Plant at 10.0–10.5 % SO_2 in the burner outlet gases. This will reduce the total volume of the gases handled and in turn will reduce the power consumption per MT of acid produced. For operating the plant at higher strengths (11–11.5 %) of SO_2 the furnace shell is air-cooled by the incoming air, which is then sent to the burner. In certain designs the furnace shell is cooled by a separate air blower and the hot air generated is used elsewhere in the premises.

11. Teflon Candle Demisters shall be provided in the DT to eliminate the possibility of acid mist carryover, which is a source of corrosion of the downstream equipment's.

12. The heat exchangers shall have Disc and Donut types of baffles instead of the usual segmental type. This will result in better gas distribution on the shell side and a lower pressure drop too.

13. Gas inlet and outlet nozzles of the equipment are designed to reduce pressure drop and for better gas distribution.

14. Gas ducting from the first pass to the second WHB can be made of SS-304 to minimise the maintenance problems due to the high temperature at the outlet of the first pass.

15. MS-PTFE lined pipelines can be used (as optional) for acid for maintenance free service as compared to C.I. pipelines in conventional Sulphuric Acid plant design.

16. Sophisticated instrumentation with DCS will be provided for the automatic control of the process parameters and for monitoring the exit gas SO_2 concentration. Data Loggers shall be provided (as an optional facility) for automatic recording of all-important process data, which can be retrieved whenever, required.

17. A very efficient Two Stage Scrubber with alkali circulation shall be installed after the FAT. A pH meter with automatic control for fresh alkali addition to the scrubber liquid will also be provided. This scrubber will be used during plant start-up/during upset process conditions only. It will not be required during steady plant operation.

18. A chimney in MS construction will be provided for the exit gases. The height of the chimney will be as per International Standards. This chimney can be mounted on FAT or on a separate foundation.

The above features are outlined in the attached block diagram (Fig. 1.1). For further information/clarification/engineering package, please contact author at Navdeep Enviro and Technical Services Pvt. Ltd, at navdeepenviro@gmail.com.

In summary, modifications to the generally practiced DCDA process are primarily the use of cesium activated catalyst an additional fifth pass to increase conversion efficiency, the use of twin oleum tower system and replacing PHE's by special alloy steel heat exchangers, an efficient acid distribution system, and PTFE lined piping for acid circulation.

Chapter 2
Chemical and Physical Properties
of Sulphur Dioxide and Sulphur Trioxide

2.1 Introduction

In order to appreciate the impact of the properties of liquid sulphur dioxide and liquid sulphur trioxide on future technology, it is important that an in-depth analysis of their properties be understood.

Though the data given in this chapter are available in literature, the practical application of the remarkable physical as well as chemical properties of sulphur dioxide and sulphur trioxide has been experienced and applied on large scale only recently.

The three main features of these two important chemicals are:

(a) High solubility of sulphur trioxide in liquid sulphur dioxide
(b) Reaction of liquid sulphur trioxide with liquid sulphur in stoichiometric proportions instantaneously to produce sulphur dioxide:

$$S + 2SO_3 = 3\ SO_2$$

(c) Liquefaction of pure sulphur dioxide at room temperatures under moderate pressures of 5–6 kg/cm^2 (Please see Fig. 2.1).

The present sulphonation techniques involves sulphonating agents such as sulphuric acid, 25 % oleum, 65 % oleum and sulphur trioxide. The technique involves high temperature reactions due to exothermic nature of sulphonation. The current techniques of sulphonation require elaborate chilling and cooling systems. Sulphonating processes currently used are generally batch operations and hence requires a battery of reactors having varying time cycles.

© The Author(s) 2016
N.G. Ashar, *Advances in Sulphonation Techniques*,
SpringerBriefs in Applied Sciences and Technology,
DOI 10.1007/978-3-319-22641-5_2

Fig. 2.1 Vaporization curves
for sulphur dioxide

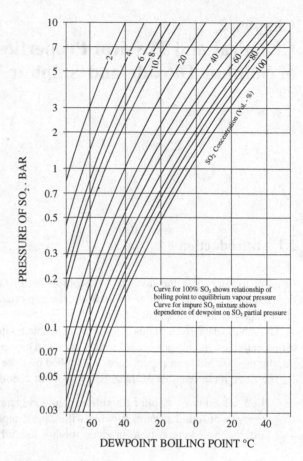

2.2   Sulphur Dioxide Physical Properties

Sulphur dioxide SO_2 is a colourless, non-inflammable, toxic gas with a characteristic pungent smell and acidic taste. Table 2.1.

2.3   Vaporisation of SO_2

It is important to analyse the physical property of condensation points at various
pressures and concentrations of SO_2.

Table 2.1 Physical
properties of sulphur dioxide

Property	Value
Molecular weight	64.06
Melting point (1013 mb)	−75.5 °C
Latent heat of fusion (at m.p)	115.6 J/g
Dynamic viscosity at 0 °C	368 Pa/s
Density at −10 °C	1.46 g/cm^3
Critical density	0.525 g/cm^3
Critical pressure	78.8 bar
Critical temperature	157.5 °C
Boiling point (1013 mb)	−10 °C
Latent heat of vaporization	
(at b.p.)	402 J/g
Standard density at 0 °C	
(1012mb)	2.93 kg/m^3
Density relative to air	
(0 °C, 1013mb)	2.263
Molar volume (0 °C, 1013 mb)	21.9 l/mol
Standard enthalpy of formation	−70.94 kcal/mol
	−4636 J/g
Specific heat, Cp (1013 mb)	
0 °C	586 K/(kg K)
100 °CC	662 J/(kg K)
300 °C	754 J/(kg K)
500 °C	816 K/(kg K)
Cp/Cv (15 °C, 1013 mb)	1.29

It can be observed from the attached Fig. 1.2 that for 100 % liquid SO$_2$ moderate pressures are required to liquefy SO$_2$ at ambient temperatures between 30 and 40 deg C.

2.4   The Solubility of SO$_2$ in Sulphuric Acid

The solubility of sulphur dioxide in Sulphuric acid (see Fig. 1.3) rises in proportion to the SO$_2$ partial pressure in good conformity with Henry's law and is increased by lowering the temperature, as represented graphically in Fig. 1.2. In the solution, sulphur dioxide is present mainly as SO$_2$ molecules, but Raman spectroscopy confirms the presence in minor proportions of the species HSO$_3$, S$_2$O$_5$ and H$_2$SO$_3$. The last of these, sulphurous acid (the anhydride of which is sulphur dioxide), exists only in aqueous solution. Aqueous solution of alkaline compounds will absorb much more sulphur dioxide than pure water (Please see Fig. 2.2) because of the formation of hydrogen sulphite (bisulphite) and sulphite ions.

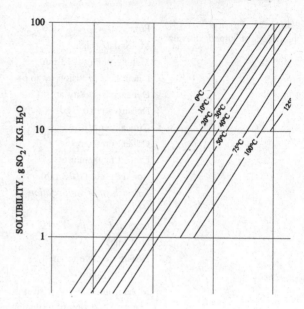

2.5   Solubility of Sulphur Dioxide in Water

It can be observed from Fig. 2.2 that the solubility of sulphur dioxide in g/kg H_2O
increases with pressure and reduces with temperature.

This property is of importance in industrial applications in scrubbling of sulphur
dioxide in tail gases.

2.6   Chemical Properties of Sulphur Dioxide

Sulphur dioxide is very stable; thermal dissociation becomes significant only above
2,000 °C. It can be decomposed by shock waves, irradiation with ultraviolet or
X-rays, or by electric discharges

The reaction of sulphur dioxide with oxygen to form sulphur trioxide is
industrially the most significant of all its reactions because of its importance in
sulphuric acid production. In the gas phase, it will only take place at elevated
temperatures and, for a satisfactory yield of sulphur trioxide; it requires the presence
of a catalyst. In aqueous solution, sulphur dioxide is oxidized to sulphuric acid at
low temperatures by air in the presence of activated coke or nitrous gases or by
oxidizing agents like hydrogen peroxide.

The reduction of sulphur dioxide with hydrogen, carbon or carbon compounds such as methane or carbon monoxide is also of industrial interest. These reactions require high temperatures or catalysts or both. They result in mixtures of elemental sulphur with hydrogen sulphide. If carbon or a carbon compound has been used as the reducing agent, carbon-containing species such as carbon dioxide, carbonyl sulphide and carbon disulphide will be formed as well.

Sulphur dioxide will normally oxidize metals at elevated temperatures, simultaneously forming metal sulphides and oxides. Liquid sulphur dioxide is a relatively efficient solvent with some water-like properties. Polar inorganic compounds are usually insoluble or only sparingly soluble in liquid sulphur dioxide, whereas covalent inorganic and organic compounds are often dissolved, mostly forming stable solutions. The fact that aromatic hydrocarbons will dissolve more readily than aliphatics in sulphur dioxide is exploited on an industrial scale for the extraction of aromatics from crude oil according to the Edeleanu process.

2.7   Physical Properties of Sulphur Trioxide

Sulphur trioxide is produced by catalytic oxidation of sulphur dioxide in concentrations of 12–15 % in gaseous form. To produce pure sulphur trioxide the plant gases are passed through oleum towers to produce 25–30 % free SO_3 oleums.

These oleums are boiled in steam heated or gas heated heat exchangers to produce pure sulphur trioxide.

This is then sent to condensers to produce liquid sulphur trioxide.

2.8   General Properties of Liquid Sulphur Trioxide

Empirical formula	SO_3
Molecular wt. of monomer	80.06
Boiling point	44.8 °C (112.6 °F)
Density (20 °C)	1.9224
Specific heat (cal/g at 25–35 °C)	0.77
Heat of dilution (cal/g)	504
Critical temperature	218.3 °C (424.9 °F)
Critical pressure	83.8 atm
Critical density	0.633 g/ml
van der Waal's constants	a = 2105
	b = 0.964

2.9   Properties of Liquid Sulphur Trioxide

See Fig. 2.3.

Property	Gamma Form	Beta Form	Alpha Form
Probable structure	$3\ S{\equiv}O \rightleftharpoons O_2S \cdots$ (ring structure)	O=S=O chains	Similar to Beta form but chains joined with one another in layered structure
Description	Liquid or Ice-like	Asbestos-like	Asbestos-like
Equilibrium melting point (°C)	16.8	32.5	62.3
Heat of fusion			
(cal/g-mole)	1800	2900	6200
(cal/g)	22	36	77
(Btu/lb-mole)	3240	5220	11,000
(Btu/lb)	40	65	139
Heat of sublimation			
(cal/g-mole)	11,900	13,000	16,300
(cal/g)	149	160	204
(Btu/lb-mole)	21,400	23,400	29,300
(Btu/lb)	268	290	367
Vapor pressure (mm of mercury)			
9° C	45	32	5.8
25° C	433	344	73
50° C	950	950	650
75° C	3000	3000	3000

Fig. 2.3   Properties of different molecular forms of liquid sulphur trioxide

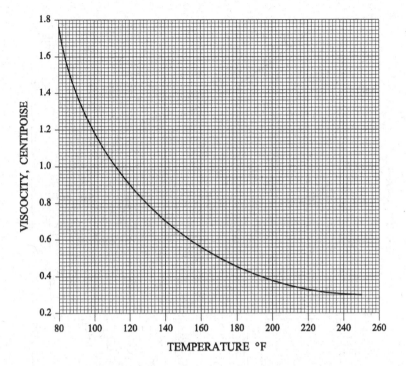

Fig. 2.4 Viscosity of liquid sulphur trioxide

2.10   Viscosity of Liquid Sulphur Trioxide

See Fig. 2.4.

2.11   Specific Gravity of Sulphur Trioxide

See Fig. 2.5.

2.12   Vapour Pressure of Liquid Sulphur Trioxide

See Fig. 2.6.

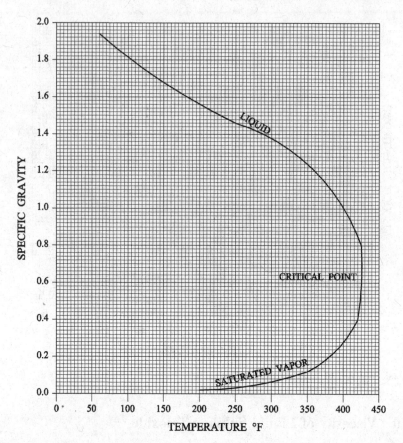

Fig. 2.5 Properties of sulphur trioxide

2.13   Molar Heat Capacity of Liquid Sulphur Trioxide

See Fig. 2.7.

2.14   Vaporisation Curves for Sulphur Dioxide

See Fig. 2.1.

2.15   Enthalpy of Sulphur Trioxide Gas

See Fig. 2.8.

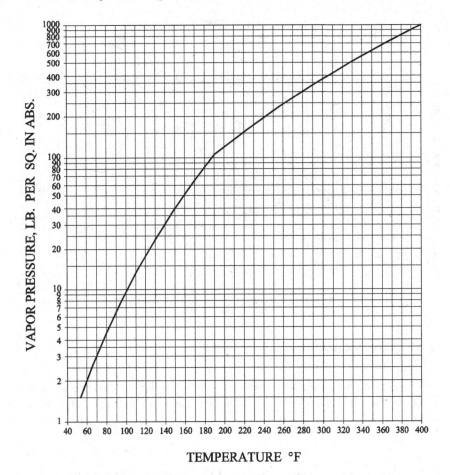

Fig. 2.6 Properties of sulphur trioxide

2.16 Chemical Properties of Sulphur Trioxide

2.16.1 *Commercially Sulphur Trioxide Is Produced by Converting 10–12 % SO$_2$ by Catalytic Conversion at Temperatures Between 360–600 °C in Multipass Converter of Sulphuric Acid Plant*

This is then further reacted with water to form Sulphuric acid by the equation
$$H_2S_2O_7 + H_2O \rightarrow 2H_2SO_4$$

It is important to note that reaction of sulphur trioxide gas with water would form micron size droplet and cannot be absorbed to form H_2SO_4.

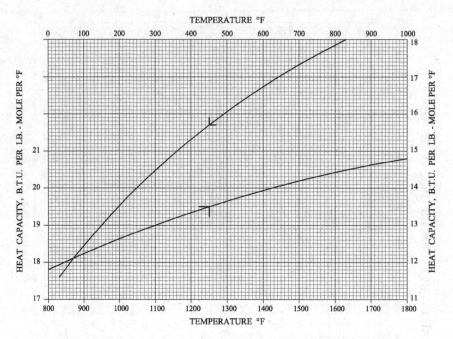

Fig. 2.7 Properties of sulphur trioxide

Formation of Sulphuric acid from SO_3 gas is exothermic and the absorbing H_2SO_4 in the absorption towers need to be cooled to maintain efficiency of absorption.

2.17   One of the Special Chemical Properties of SO_3 Which Has Been Safer but not Explored till date

$$S + 2SO_{3liq} \rightarrow 3SO_{2gas}$$

$$\Delta H = -74.3 \text{ kcal/g mole}$$

$$\Delta F = -36.71 \text{ kcal/g mole}$$

Since the free energy change is large and negative, the reaction is almost instantenous. In addition, the reaction generates one additional mole in gaseous form, so there is a pressure increase.

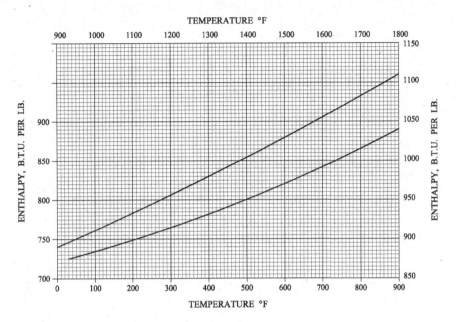

Fig. 2.8 Properties of sulphur trioxide

2.18   Sulphur Trioxide Is a Strong Sulphonating Agent for Difficult, Organic and Inorganic Chemicals

2.18.1   Treatment of Sulphuric Acid Plant Tail Gas from Final Absorption Tower

The tail gases of sulphuric acid contact plants consist chiefly of nitrogen and residual oxygen. They also contain sulphur dioxide in low concentrations which depend on the conversion efficiency attained in the conversion stages. The content of gaseous sulphur trioxide and sulphuric acid is essentially a function of the temperature and concentration of the irrigation acid in the final absorber. Under unfavourable operating conditions, as, for example, when the sulphur dioxide-containing converter feed gases are inadequately dried or contain hydrocarbons, sulphuric acid mists can be formed which are not removed in the absorption system, even when the concentration and temperature of the absorber acid are at their optimum values. The safest way of removing these acid mists is using a candle type demister. However, this is not very effective in removing excessive sulphur trioxide concentrations, which may result from poor acid distribution in the absorber.

Chapter 3
Manufacture of Sulphonating Agents Such as 25 and 65 % Oleums as well as Liquid Sulphur Trioxide

3.1 Introduction

In order to produce gaseous as well as liquid trioxide it is essential that manufacture of oleums by absorption of sulphur trioxide from plant gases producing sulphuric acid.

Pure (99 % +) sulphur trioxide can be produced by boiling 25 % oleum. This can be liquefied or used for producing 65 % oleum.

3.2 Production of 25 % Oleum

3.2.1 History

Until the mid-50s all sulphonations were carried out by use of Sulphuric Acid of different strengths. The reactions carried out by Sulphuric Acid were not able to give the conversion efficiencies that were economical, especially in the manufacture of detergents. In certain dye intermediates the higher strength Sulphuric Acid was found to be an essential part of the chemical reaction. This led to many of the Sulphuric Acid plants to manufacture "Fuming Sulphuric Acid" popularly known as Oleums.

© The Author(s) 2016
N.G. Ashar, *Advances in Sulphonation Techniques*,
SpringerBriefs in Applied Sciences and Technology,
DOI 10.1007/978-3-319-22641-5_3

3.3   Technical Considerations

The initial product was in the range of 20–25 % Oleum which meant 20–25 % free SO_3 dissolved in Sulphuric Acid.

The basic reaction carried out in the Oleum Tower is absorption of SO_3 in the Sulphuric Acid to form an adduct $H_2SO_4 + SO_3 = H_2S_2O_7$

This would correspond to 45 % free SO_3 dissolved in Sulphuric Acid. Since the freezing point of 40–45 % oleum is in the range of 90–95 °F (35–40 °C), there is always a possibility of solidification and consequent choking of equipment and pipelines at ambient temperatures, especially in the cold season (See Fig. 3.1).

Hence, the industrial preference for manufacturing for fuming Sulphuric Acid was in the range of 20–25 % free SO_3. This needs removal of exothermic heat of dilution together with heat of absorption of SO_3 by specially designed coolers. Some plants have a gas heated oleum boiler for boiling the 25 % oleum to produce pure SO_3 vapour.

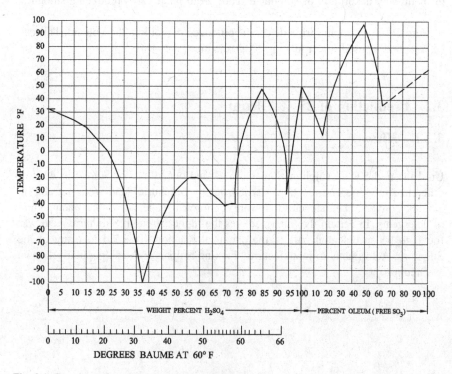

Fig. 3.1 Freezing point of H_2SO_4, oleum and SO_3

3.4 Manufacturing

The manufacture of Oleum is generally in accordance with the standard Flow Diagram (see Pictorial view # 1 & # 2). The plant gases from the Contact Sulphuric Acid plant (based on sulphur) having SO_3 strength of 10–12 % are first cooled by counter current heat in the Cold Heat Exchanger (CHE) after Pass-III. The temperature of gases leaving cold heat exchanger (CHE) will be in the range of 270–280 °C and the same is further cooled by an Economiser for pre-heating the Boiler Feed Water. The cold gases in the range of 140–160 °C from the economiser outlet are fed to the Oleum Tower.

The strength of the Oleum is automatically controlled by an indicator-cum-controller to control the flow of Sulphuric Acid for dilution. This addition is done through an Absorption Tower so as to prevent any fumes of SO_3 gases from the Oleum Tower System. The level is controlled through a Level Controller allowing oleum to be pumped to the storage for cooling.

In addition to the exothermic heat generated in the Oleum Tower, there is a sizeable sensible heat from the plant gases transferred to the Oleum since the equilibrium demands the operation of the Oleum Tower in the range of 50–60 °C. Earlier, this cooling was done by Trombone Coolers with external water spray made from Shcdule-80 M.S. pipes. However, the compact Plate-type Heat Exchangers now available are widely used by the industry. Since the absorption of SO_3 in the Oleum requires very close control of temperature, it is important that the cooling of Oleum is done by cooling water supplied from a dedicated Cooling Tower for the 25 % oleum section. Alternatively, for cooling of oleums, special coolers in SS-Alloy construction are available with anodic protection. These are selected purely from economic considerations.

The cold Oleum is then despatched to the consumers in tankers having capacities varying from 10–20 MT. For special customers, SS-316 tankers are used instead of MS tankers.

3.5 65 % Oleum

3.5.1 Introduction

The demand for higher strength Oleums has made it imperative to go for those strengths of Oleum, which are commercially viable and transportable. As it can be observed in Fig. 3.1, the strengths between 30–50 % are not practical in view of the high freezing points. Hence, it was considered practical to manufacture Oleum having free SO_3 content of 60–65 %.

3.6   Manufacturing

Initially, 65 % Oleum was manufactured by addition of One tonne of Liquid SO_3 in One tonne of 28–30 % Oleum in a mixer with cooling coil arrangement. This was made in a batch manner and the product was then transferred to the storage for sale.

Unlike 25 % Oleum, 65 % Oleum can be manufactured only by use of pure SO_3 gases evaporated out of 25 % Oleum in a boiler heated by energy available in the hot gases of the Contact Sulphuric Acid Plant. Please refer to Block Diagram (see Fig. 2.8). The pure SO_3 vapours are sent to 65 % Oleum Tower having facilities for cooling the heat generated due to exothermic absorption as well as dilution by addition of 25 % Oleum. A Level Controller is provided to maintain Tank level in the circulation boot.

The product from the continuous plant manufacturing 65 % Oleum is sent to the storage for despatch to the consumers in tankers having capacity 10–20 tonnes. 65 % Oleum is transported in SS-316L tankers instead of MS when the iron content in the product is to keep low.

3.7   Uses

The demand for 65 % Oleum is primarily for the manufacture of special dyes and intermediates such as H-Acid. As a rule of thumb, one tonne of H-Acid requires 3 tonnes of 65 % Oleum.

65 % Oleum is also used directly in other Sulphonation reactions.

As a raw material for generating pure SO_3 gas for Sulphonation was considered as hazardous and difficult.

3.8   Sulphur Trioxide (Liquid or Gas)

3.8.1   Introduction

25 % Oleum and 65 % Oleum were introduced in chemical processing in search of stronger Sulphonating agents. However, with the sulphonation process, SO_3' was used and by-product weak Sulphuric Acid needed disposal as waste product. Environmental consideration led to large Effluent Treatment Plants (ETPs) with consumption of *lime* in large quantities. With the result that heaps of waste gypsum were formed and posed solid waste disposal problem.

Due to environment and economic considerations, uses of pure sulphur trioxide in gaseous or liquid form were investigated. The main consideration was to tame the exothermic reaction and avoid unwanted co-products.

In 1960 in India, the production of Liquid Sulphur Trioxide was less than 100 tonnes per annum (TPA). Today there are plants in India, which produce more than 100 TPD of liquid or gaseous sulphur trioxide with purity of 99–99.5 %. The total production of pure sulphur trioxide in India can be estimated as about 300,000 MTs in the year 2003. 60 % of the production is for captive use.

3.9 Manufacture

The manufacture of liquid or gaseous SO_3 is carried out by boiling of 25–32 % Oleum as in the case of manufacturing 65 % Oleum. The sulphur trioxide can be liquefied by condensing at 43–45 °C. Care should be taken to prevent any moisture ingress in the plant as it may cause serious operating problems.

The liquefied SO_3 is then taken to a product storage tank from which tankers are filled using Intermediate Lorry Filing Tanks.

3.10 Economic Considerations

For any Sulphuric Acid plant, the production of liquid SO_3 is more profitable and hence the objective is to maximise the production of SO_3. Two important considerations for this are as follows:

a. **Mass transfer efficiency**

Absorption of SO_3 from converter gases in oleum tower—higher efficiency can be obtained when strength of circulating oleum is lower. But this can lead to lower production rate of SO_3 vapours from the GHOB/SHOB. Hence, a latest innovation to maximise the production of SO_3 is to adopt a Twin Tower System wherein two Oleum Towers operate in series with lean and strong Oleum. This is found to be cost-effective and enhances overall efficiency above 40 %.

b. **Evaporation efficiency**

The circulating oleum is to be boiled by the heating medium (either plant gases for steam) to produce SO_3 vapours. If the boiling is continued for longer time, the amount of SO_3 vapour produced per metric tonne of oleum (fed to the boiler) will be more. However, this can result in a lower strength of the oleum leaving the boiler. Since oleums of strength below 20 % are very corrosive, this situation is to be avoided. Hence, from the practical point of view, the strengths of oleums are 28–32 % at inlet to the boiler and 22–23 % at exit from the boiler. (Please see pictorial # 2 and Fig. 1.1).

The lean oleum from the exit exchanges heat with the incoming strong oleum for preheating before entry into the boiler. It is then further cooled in separate lean oleum coolers before putting it back into the oleum tower for fortification by plant gases.

In a twin-tower system, the first oleum tower operates at 30–35 % oleum strength while the second oleum tower, which is in series with the first one, operates at 21–24 % for further absorption of SO_3 (from the exit gases of the first oleum tower).

Chapter 4
Manufacture of Liquid Sulphur Dioxide

4.1 Manufacture of Liquid Sulphur Dioxide

The conventional method of producing sulphur dioxide is by combustion of sulphur followed by absorption-desorption, drying and cryogenic condensation. This process is capital intensive, hazardous, saddled with corrosion and high utility consumption.

$$S + O_2 \rightarrow SO_2 \quad \Delta H = -297 \text{ KJ/Mol}$$

Molten sulphur at 140 °C is fired in a 8–10 meter long furnace using preheated (80 °C) air. The theoretical temperature achieved by the exit gases is given in Fig. 4.1. For the production of liquid SO_2 high concentrations 16–18 % are necessary. Since the adiabatic temperature in single stage combustion will be high (1600–1800 °C), a two-stage combustion system can be used. (Please see Fig. 4.1)

Subsequently pure sulphur dioxide is produced by chemical reduction of sulphur trioxide with sulphur. This route is more direct and attractive due to the fact that sulphur trioxide is produced in a Sulphuric Acid plant. However, hitherto the process used by IG Farben Industrie of Germany under US patents 3,432,263 and 3,455,652 (1966) is from solid or molten sulphur to 25 % Oleum at 110 °C giving 98.5 % SO_2 and 1.5 % SO_3. The excess sulphur trioxide is further reacted with solid sulphur in a tower and then traces of sulphur trioxide are removed by absorbing in 98 % Sulphuric Acid prior to compression and condensation.

NEAT's process deals with use of liquid sulphur and liquid sulphur trioxide under pressure (8 to 10 kg/cm^2) to produce pure sulphur dioxide at relatively low temperature without need for compression or refrigeration.

© The Author(s) 2016
N.G. Ashar, *Advances in Sulphonation Techniques*,
SpringerBriefs in Applied Sciences and Technology,
DOI 10.1007/978-3-319-22641-5_4

Fig. 4.1 Theoretical sulphur combustion temp as a function of combustion gas SO_2 concentration inlet air at 80 °C, t(sulphur)

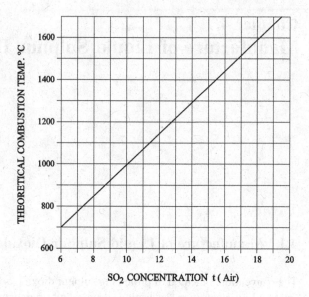

4.2   Thermodynamic and Kinetic Consideration of the NEAT's Process

It is interesting to note that a highly exothermic nature of sulphur oxidation in a furnace at about 1000 °C can be carried out at reasonably low temperatures 50–110 °C in a pressurised reactor. This is possible by reaction.

$$S(Liq) + 2\ SO_3(Liq) \rightarrow 3SO_2(Gas)$$

$$\Delta H = -74.3\ KCal/g\ mole$$

$$\Delta F = -36.71\ KCal/g\ mole$$

Since the free energy change is large and negative, the reaction is almost instantaneous. In addition, the reaction generates one additional mole in gaseous form. Thus the pressure of the reactor builds up.

If there is a stoichiometric addition of liquid sulphur trioxide (30–40 °C) and molten sulphur (135–140 °C), under a pressure of 8–10 kg/cm^2, then the sulphur dioxide (after cleaning) formed (98–99 %) can be liquefied at room temperature. The exothermic heat removal of about 15625 kCal/ton of sulphur dioxide produced is achieved by circulating cold water through the jacket.

Since the reaction is instantaneous and complete, the reactor volume is very small. However, oleum 25 % is used as a carrier to provide uniform mixing of liquid sulphur trioxide and liquid sulphur. Agitation with proper mechanical seal is provided to ensure complete reaction.

The increase in molecules of SO_2 by fifty percent enables build-up of pressure without use of compressor. Figure 2.1 gives condensation temperature of pure sulphur dioxide at different pressures. This indicates that if a pressure of 7 to 8 kg/cm^2 is maintained in the system, the liquefaction can be done at ambient temperature using water at 35–40 °C, cooling is required at the rate of 1.12 million kCal/ton of product.

The process for the above is shown in block diagram. (Please see Fig. 4.2)

4.3   International Scenario

Fifty years ago the world production of liquid sulphur dioxide was less than 1,00,000 tons. Over the past few decades use of sulphur dioxide in petroleum refining as solvent, in manufacture of paper pulp, in textiles as sulphites, bisulphites and hydrosulphites and in effective control of fermentation in wine making and as preservative for fresh fruits and vegetable has increased the demand many fold.

END USES…Sulphur dioxide, sulphurous acid anhydride, is a non-flammable, colourless gas at normal ambient temperature with a characteristic pungent odour. It is supplied as a water white clear liquid-compressed gas with a purity of 99.98 %. Commercial grade containing not more than 0.05 % moisture is suitable for most applications. It has a boiling point of 14 °F(-10 °C).

Applications of sulphur dioxide are diverse. It is used:

- To make other chemicals such as bisulphides, metabisulphites, thiosulphides, sulphites, hydrosulphites, and sulphonates.
- To produce dimethyl sulphoxide or thionyl chloride.
- Directly in sulhpite pulping in paper industry
- In the production of in situ sodium hydrosulphite,
- As a redundant in the production of chlorine dioxide from sodium chlorate, and
- To remove excess hydrogen peroxide in the bleaching process.
- In food and agriculture
- In corn processing to remove the kernel hull for making high fructose corn syrup and ethanol.
- As a sterilant, preservative and bleach in certain food and beverage products. in water treatment it is used as a chlorine scavenger, reducing free chlorine
- In waste-water treatment plant discharges.
- In metallurgical processing, SO_2 is used in the purification of certain elements from their ores, for the recovery of certain elements from mixtures of other materials, and increasingly in the reduction of cyanides in leachate from gold mining. In pollution control to reduce hexavalent chromium ions to be more innocent trivalent form for easier disposal.

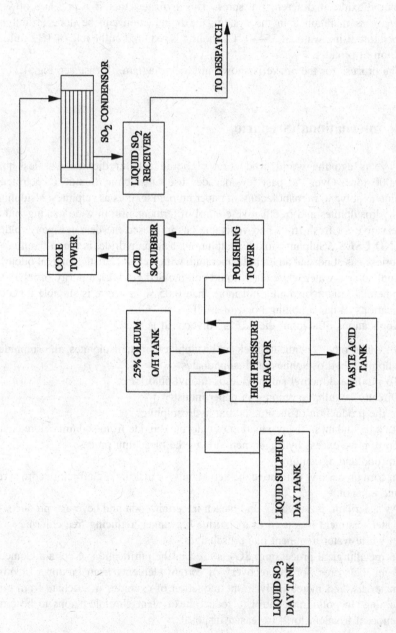

Fig. 4.2 Block diagram for liquid SO₂ plant

4.4 Merchant Market for SO_2 in Various for Many Industrial Applications

Based on an international survey, Table 4.1 describes the percent usage for different chemicals

4.5 Process Description

Continuous production of liquid Sulphur dioxide using liquid Sulphur and liquid Sulphur trioxide is economically feasible when it is attached to an existing Sulphuric Acid plant producing oleums and liquid SO_3. A block diagram is given in Fig. 4.2:

Sulphur trioxide gas is received from the main Sulphuric Acid plant by boiling oleum. Hot Sulphur trioxide is cooled and condensed in SO_3 Condenser and stored in SO_3 Storage Tank. Likewise, liquid Sulphur is received in a Day Tank. Molten sulphur and liquid SO_3 are pumped in the reactor at regulated rates using metering pumps. The reactor is initially charged with 25 % oleum from the overhead tank.

The temperature of the reaction is maintained by circulation of hot/cold water using vessels and pumps respectively. The sulphur dioxide gases are then reacted to polishing towers in which lumps of solid sulphur react with the unreacted Sulphur trioxide. The gases then subsequently pass through a Sparger containing Sulphuric Acid for removal of trace sulphur trioxide. The gases then pass through a Demister and a coke tower to remove the acid mist.

Pure sulphur dioxide is then condensed in condensers and then received in receivers. These are then sent for loading tankers from which one tonner cylinders or road tankers are filled for sale. (Please see Pictorial view #4 & #5) of 15 TPD liquid SO_2 plant under construction at ATUL/AMAL (Ankleshwar, Gujarat) in 2004.

Table 4.1 International percentwise use of sulphamic acid in various industries

Derivative	Percent	Derivative	Percent
		Agriculture & Food (incl. HFCS)	15
Sodium Sulphite (SS), Sodium Bi Sulphite (SBS), Sodium meta bisulphite (SMBS), Hydrosulphite	20 20	Water Treatment Miscellaneous	10 10
Pulp and paper	25		

4.6   Operational Considerations

4.6.1   Condensation and Filling Section

It is important that all the gases of SO_3 are absorbed in the Sulphuric Acid Sparger and the Sulphuric Acid mist is removed by cyclone separator as well as mist eliminator and traces are trapped in Coke Tower.

Care should be taken in handling liquid SO_2 under pressure, and protective gear must be available for handling liquid SO_2 leakage or SO_2 vapours.

The clean SO_2 gas under pressure between 7 and 8 kg/cm^2 enters the Condenser through a non-return valve and is condensed by water from the Cooling Tower. The quantity being condensed to be seen through look glass and will be stored alternatively in liquid SO_2 receivers from this it will go to the Filling tank for filling the cylinders together with facility for Weighing Machine generally as per the normal practice of handling liquefied chemical gases under pressure. A scrubbing system for purge is provided using alkali.

Safe Operation of the liquid SO_2 plant can be ensured by following simple precautions as given below:

- At any time do not allow accumulation of sulphur in the reactor
- At any time do not allow the temperature of the reactor go beyond 70 °C
- The purge system must be tested at least once a day.
- Do not allow ingress of water in any section of the plant giving any contact with SO_2/SO_3.
- Keep an eye on the pressures, temperatures and ampere of the electric drives.

4.7   Economics

Hitherto, liquid SO_2 has been manufactured by compression and refrigeration using furnace gases by burning sulphur with air in a furnace or by using oleum and sulphur.

Economically both of the above technologies have inherent handicap. The capital cost per ton of liquid SO_2 in the above technologies will be at least 2–3 times higher than the capital cost required for the process recommended in this chapter.

As regards raw material and utilities, in both processes one third cost is common by using sulphur. In the process using 25 % oleum there is a continuous bleed of oleum producing 99.5 % Sulphuric Acid, which may have low resale value because of its colour, and that it can only be used in the manufacture of Phosphatic fertilisers. Whereas in the process recommended in this chapter, the liquid SO_3 will be completely reacted, and only occasionally the reactor have to be drained for removal of unreacted impurities in the sulphur. The cost of sulphur trioxide, which will be higher than oleum, is to be balanced against the saving on interests, depreciation and maintenance of capital equipment as well as saving of utilities. This cost can be reduced to the cost of sulphur by a novel process under development by NEAT.

4.8   Environmental Considerations

Most of the Sulphuric Acid plants world over have the problem of disposal of settled sulphur sludge. Modern plants have a sulphur filter, and the filter cake containing 50–70 % sulphur is a sludge disposal problem. This can be converted to SO_2 and subsequently converted to 98 % Sulphuric Acid by the process described in this process with minor modifications. NEAT has applied for an international patent for the same.

From environmental considerations, the storage of sulphur cake is a fire hazard and a solid waste disposal problem. The *leachate* being acidic creates liquid effluent problems.

It is suggested that the sulphur sludge in a slurry/molten form can be pumped into the reactor and be converted to sulphur dioxide. In case if there is no market for liquid sulphur dioxide, the reactor can operate at moderate pressures and recycle the sulphur dioxide thus produced to the Sulphuric Acid plant for conversion to SO_3 and Sulphuric Acid.

4.9   Conclusion

The use of liquid sulphur and liquid SO_3 under pressure fed into a reactor to produce liquid sulphur dioxide has been found to be the most environment friendly safe and economical option.

Chapter 5
World Production of Liquid SO_2 and SO_3

5.1 Introduction

The world production of liquid SO_2 and liquid SO_3 has increased considerably due to the production of speciality chemicals required for petroleum, pharmaceutical and dye industries.

The percentage distribution of demand is indicated in the table in Chap. 4. As given in Tables 5.1 and 5.2, this is likely to change due to large-scale demands for petroleum, pharmaceuticals and dye industries.

5.2 World Scenario

The current status of world capacity and production of liquid SO_2 and liquid SO_3 are given as under:

World Scenario the data available with us is for year 2010 as follows:-Tables 5.1 and 5.2

5.2.1 Comparative Analysis on Techno Economic Considerations

Manufacture of liquid SO_2 has gone through a radical change to reduce raw material and utility costs. Comparative production pattern indicates the closing down of liquid SO_2 plants which were uneconomical and growth of plant capacity

© The Author(s) 2016
N.G. Ashar, *Advances in Sulphonation Techniques*,
SpringerBriefs in Applied Sciences and Technology,
DOI 10.1007/978-3-319-22641-5_5

Table 5.1 Sulphur di-oxide producers (Merchant) and Capacities—(thousands tons)

	Producer	Location	1999 capacity	2010 capacity
USA	Calabrian Corp	Port Neches TX	50	180
	Clariant Corp	Bucks, AL	65	NIL
	Marsulex	Copperhill, TN	45	Shutdown
	Marsulex	Cairo, OH	13	20
	PVS Chemical	Chicago, IL	40	36.5
	Republic Group (Olin)	Charleston, TN	45	ND
	Rhodia	Baton Rough, LA	25	DA
	Rhodia	Houston, TX	43	DA
	Thatcher Company	Salt Lake City, UT	10	ND
		Total USA	399	
Canada	Cominco	Trail, BC	88	88
	Marsulex	Timmons, ON	33	Shutdown
	Marsulex	Prince George,BC	33	Shutdown
	Marsulex	Sudbury, ON	110	110
		Total Canada	264	198
India	Transpeck (Batch Process)	Baroda	12	12
	Shree Sulphuric Acid (NEAT)	Ankleshwar	0	06
	Atul Ltd (NEAT)	Ankleshwar	0	06
	Nath Industries (NEAT)	Vapi	0	06
	Aarti Ltd (NEAT)	Vapi	0	06
		Total India	12	36

DA Data Awaited
ND No Data

using direct process of reacting under pressure liquid sulphur and liquid SO$_3$ under stoichiometric proportion by the reaction

$$S + 2SO_3 = 3SO_2 \Delta H = -99\,KJ\,mole$$

As regards raw material and utility consumption by the process adopted by NEAT, the following are guaranteed figures per tonne of liquid SO$_2$

Liquid SO$_3$	0.841 MT
Liquid Sulphur	0.175 MT
Power	30 KWH
Water (cooling)	1.5 M^3
Steam (low pressure)	0.3 MT

Table 5.2 Liquid SO$_2$ plants in Europe Africa and South America

Company	Location	Capacity	Year	Source
Boliden Harjavalta Oy	Harjavalta, Finland	42,500 MTPA	2010	ND
Boliden Mineral AB	Skelleftehamn, Sweden	50,000 MTPA	2010	Water absorption/stripping followed by drying and liquefaction
Calabrian Corporation	Port Neches, Texas, USA	180,000	2010	ND
Chemtrade Logistics	Cairo, Ohio, USA	20,000 STPA	2010	React sulphur and sulphur trioxide
Goro Nickel S.A.S.	New Caledonia	2 × 25 MTPD	2010	ND
Grillo-Werke AG	Duisburg, Germany	ND	ND	Absorption Process
Grillo-Werke AG	Frankfurt/Main, Germany	ND	ND	Condensation process
Marsulex Inc.	Prince George, British Columbia, Canada	90 MTPD	1989	Partial liquefaction of high strength SO$_2$ gas from sulphur burning Shut down in 2010
PVS Chemical Solutions Inc.	Chicago, Illinois, USA	36,500 STPA	2010	ND
Sable Zinc Kabwe	Kabwe, Zambia	6 MTPD	2007	ND
Teck Cominco	Trail, British Columbia, Canada	88,000 MTPA	1999	Acidulation of ammonia from ammonia scrubber bleed
Vale INCO	Sudbury, Ontario, Canada	400 MTPD	1999	Compression and liquefaction of high strength metallurgical gases from Inco flash furnace
Votorantim Metais	Juiz de Fora, Brazil	35 MTPD	1999	ND
Xstrata Copper	Timmins, Ontario, Canada	ND	2010	Partial liquefaction of high strength SO$_2$ gas from metallurgical gases
Xstrata Zinc —Hinojedo Roasting Plant	Spain	ND	ND	Liquefaction

ND No Data

Any other process currently in practice requires higher utility costs. As regards raw material, equivalent sulphur required is 40 % of 0.841 = 0.3364 plus 0.175 i.e. 0.511. Part of purged SO_2/SO_3 is returned to Sulphuric acid plant. Hence, Sulphur efficiency is higher than 98 percent.

Besides, the capital cost is very low since the material of construction is Mild Steel. Also, maintenance cost is low due to lack of compression or refrigeration. The plant is more safe and environment friendly.

5.3 Economics of Manufacture of Liquid SO₂

Hitherto, liquid SO_2 has been manufactured by compression and refrigeration using furnace gases by burning sulphur with air in a furnace or by using oleum and sulphur.

Economically both of the above technologies have inherent handicap. The capital cost per ton of liquid SO_2 in the above technologies will be at least 2–3 times higher than the capital cost required for the process recommended in this chapter.

As regards raw material and utilities in both processes one third of the cost is common becuase of the use of sulphur. In the process using 25 % oleum, is continuous bleed of oleum producing 99.5 % Sulphuric Acid which may have low resale value because of its colour and can be utilised in the manufacture of Phosphatic fertilisers only. Whereas in the process recommended in this document the liquid SO_3 will be completely reacted and only occasionally the reactor will have to be drained for removal of unreacted impurities in the sulphur. The cost of sulphur trioxide which will be higher than oleum is to be balanced against the saving on interests, depreciation and maintenance of capital equipment as well as saving of utilities. This cost can be reduced to the cost of sulphur by a novel process under development by NEAT.

Interest on capital employed: The initial investment will cover plant and machinery, civil structures (including administration building, stores and maintenance shed), infrastructure for water treatment, storage tanks for products, storage yard for raw sulphur, road for internal movement, storm water drains, effluent treatment plant etc. The capital investment will thus depend on: (a) production capacity planned for each of the products for the present and the future; (b) the infrastructure required at present and for the future.

5.4 Economics of Manufacture of Liquid SO₃

A Sulphuric Acid plant generally produces Technical grade acid. However, the plant management should explore the markets for the following products also since they can realise much higher prices, and can be produced by "adding on" certain equipments as outlined below. This will improve the financial profitability of the organisation.

Product Mix

- Technical grade Sulphuric Acid
- Battery grade/Analytical Reagent grade Sulphuric Acid: This will need PTFE lined/glass absorption tower, special PTFE lined pumps, glass heat exchangers etc.
- Electronic grade for manufacture of electronic Printed Circuit Boards.
- 23–25 % Oleum: An additional absorption tower with Oleum circulation pump, trombone type or PHE type Oleum Coolers, dedicated cooling tower etc.
- 65 % Oleum: In addition to the equipments required for the manufacture of 23–25 % Oleum, this will need a separate absorption tower, Oleum circulation pump, cooling system etc. and facilities for generation of pure SO_3 vapours (by boiling 25 % Oleum). Certain plants produce 65 % Oleum by mixing 30 % Oleum with liquid sulphur trioxide in a controlled manner.
- Liquid Sulphur Trioxide (SO_3): Needs specially designed water-cooled Condenser, a dedicated cooling tower, storage tanks etc.
- **Liquid Sulphur Dioxide (SO_2)**: Conventional processes for manufacture of liquid sulphur dioxide are based on burning of sulphur to produce gases with 16–18 % SO_2, absorbing the SO_2 in water, stripping it to give moist SO_2 which is, in turn, dried, compressed and subsequently liquefied by refrigeration. This needs considerable amount of energy for compression and refrigeration. An innovative process developed and put into operation by NEAT Services Pvt. Ltd., India is based on direct reaction of elemental sulphur with liquid sulphur trioxide in a specially designed reactor to produce pure dry sulphur dioxide, which can be liquefied by ordinary cooling tower water. This process does not need compression and refrigeration and hence the cost of production of liquid sulphur dioxide is considerably less than the conventional process. Four plants based on this process have already been in operation in India for the last 10 years and above. One more plant is under erection having 15 TPD capacity by Addar Group of Industries at Riyadh, Kingdom of Saudi Arabia. It is expected to be commissioned by early 2016.
- **SO_3 bearing gases for sulphonation**: About 90–95 % of the SO_2 fed to the converter is converted to SO_3 by the end of the third-pass. The gases coming out from the exit of the third-pass thus contains up to 10–15 % SO_3 and hence can be used for sulphonation of Linear Alkyl Benzene (LAB) for the detergent industry by providing some additional equipments. The unconverted SO_2 can be scrubbed with alkali scrubber after the sulphonation plant.

5.5   Conclusion

The above indicates that the most economical choice for making liquid SO_3 is the twin-tower system and for liquid SO_2 the process developed by NEAT[1] of high pressure reaction of liquid SO_3 and liquid sulphur are the most economically viable options.

It is important to note that the above processes are safe and environment friendly.

---

[1]Navdeep Enviro and Technical Services Pvt. Ltd, **Email**: navdeepenviro@gmail.com.

Chapter 6
Techno Economic Evaluation of Processes Involved to Manufacture Liquid Sulphur Dioxide and Liquid Sulphur Trioxide

6.1 Introduction

Sulphur dioxide has been a key chemical starting point of many chemical products including Sulphuric Acid, liquid sulphur dioxide, oleums, liquid sulphur trioxide etc., Recently the requirement for liquid sulpur dioxide has increased due to production of specialty chemicals in the production of petroleum products.

With the uncertainty of the prices of oil, this industry has become very competitive. Hence, technologies established over several decades have to be revalued for their capital cost, utility cost, manpower requirements, maintenance, environment and safety considerations.

It is also the starting step for the cold process for the manufacturer of sulphuric acid, liquid sulphur trioxide, oleum under high pressure conversion with zero emission of sulphur dioxide.

Industrially Sulphur dioxide is produced in greater quantity than any other Sulphur compound for the past century. The major derivative of Sulphur dioxide is Sulphuric acid which is considered as a barometer of chemical industry of any country. That is why Sulphur dioxide—the starting material of Sulphuric acid and Sulphur derivatives is known as the 'King of Chemicals'.

Sulphur dioxide is generated by combustion of elemental Sulphur or by oxidation of Sulphides. Very large volumes of Sulphur dioxide are produced involuntarily by roasting of Sulphides to extract metals and also by combustion of Sulphur bearing hydrocarbons.

Sulphur dioxide has been known since ancient times for its disinfecting and bleaching quality. Since the late 19th century and early 20th century, it has been used for industrial refrigeration plants. Also, it is widely used for producing quality paper by preparing Sulphite pulp. Quality pulp produced is taken as a raw material

© The Author(s) 2016
N.G. Ashar, *Advances in Sulphonation Techniques*,
SpringerBriefs in Applied Sciences and Technology,
DOI 10.1007/978-3-319-22641-5_6

for producing viscose rayon. Lately liquid Sulphur dioxide has become a major component in petroleum refining and sulphonation reactions.

6.2   History

Sulphur dioxide on industrial scale has been manufactured in the early twentieth century. The advent of contact process in the thirties opened the door for new techniques to produce pure Sulphur dioxide. In the late eighties, due to environmental considerations, Sulphur recovery from crude oil and flue gas cleaning from combustion of Sulphurous fossil fuels added to the SO_2 recovery system.

Following are the main sources of Sulphur dioxide production:-

1. Elemental Sulphur
2. Pyrite
3. Sulphide ores of non-ferrous metals
4. Waste Sulphuric acid and Sulphates
5. Gypsum and anhydrite
6. Hydrogen Sulphide-containing waste gases
7. Flue gases from the combustion of Sulphureous fossil fuels
8. Destructive combustion of gypsum and other waste Sulphuric acid containing inorganic or organic salts.

6.3   Production by Burning Sulphur Cooling, Absorption in Alkali and Desorption, Drying by Sulphuric Acid, Compression and Condensation by Refrigeration

In the late fifties, it was used by those plants which were dedicated totally to produce only liquid Sulphur dioxide. Apart from high cost of the plant, this process required special materials of construction like Hastelloy—C due to heavy corrosion. However, for the production of alkali sulphites and bi-sulphites, this process is still in operation in certain plants.

This process is still in use for production of Sulphur dioxide in Canada for metallurgical plants.

Due to high capital cost as well as maintenance cost, this process is now being replaced by direct reaction of liquid Sulphur with liquid SO_3 as described in this document.

6.4   Production by Use of Organic Solvent from by Product SO$_2$ Generated in Specific Chemical Reactions

By product Sulphur dioxide is produced in the manufacture of petroleum additives, Thionyl chloride, continuous detergent manufacture, etc.

Sulphur dioxide dissolves readily in most organic liquids such as methanol, ethanol, benzene, acetone and carbon tetrachloride. Sulphur dioxide is completely miscible even at low temperature with ether, carbon di sulphite, chloroform and glycol.

By product Sulphur dioxide is absorbed in any of the solvents and then released as pure Sulphur dioxide recycling the solvents for absorption.

Depending on pressure at which liquid SO$_2$ is released from the solvent, the condensation takes place as per Fig. 1.2. It can be observed that if the pressure of desorption tower is maintained between 7–10 kg/cm^2, no refrigeration will be required.

Large quantities of liquid SO$_2$ are produced by this process as by product.

6.5   Production by Use of Concentrated Oleum (65 %) and Solid Sulphur Using Compression and Refrigeration (Batch Process)

This process is a batch process which was adopted for the production of small quantities of liquid SO$_2$.

Solid Sulphur is charged in a reactor fitted with agitator and cooling jacket, concentrated Oleum is introduced in stochimetric proportion to produce concentrated Sulphur dioxide (97.5 %) and free SO$_3$ (2.5 %). The gases are led to a Sulphuric acid tower to remove unreacted Sulphur trioxide after passing through demister. The pure SO$_2$ is compressed and chilled to give the final product.

This process is still in use in India in some of existing manufacturing units. However, the cost of production being high these are being replaced by latest technique of producing liquid Sulphur dioxide using molten Sulphur and liquid SO$_3$.

6.6   Production by Using Molten Sulphur and Liquid SO$_3$ Under Pressure Without Compression and Refrigeration (Adopted by NEAT)

This process is described in detail by the author in Reference 5, paper presented at British Sulphur conference in 1999 at Calgary-Alberta.

Currently in India, more than 24,000 TPY of liquid SO_2 is being produced by the 'NEAT' process.

Currently NEAT is in the process of putting up 4000 TPY liquid SO_2 plant in Saudi Arabia.

As regards the economics for the various plants, it is described in the next section.

6.7   Economic Considerations

The detailed economic evaluation of the processes involved has already been evaluated in Chap. 5.

For the manufacture of SO_3 using cold process and manufacture of SO_2 using high pressure reactor fed with metered quantities of liquid sulphur and liquid SO_3 are economically the most viable options.

For a 4000 TPY

Part A—For 15 TPD liquid SO_2 plant Energy Balance and Material balance in Appendices section.

Part B—For 12 TPD liquid SO_3 plant material and energy balance in Appendices section.

6.8   Conclusion

The above analysis clearly indicates the future of Sulphuric acid and sulphonating agents technology which can be applied in an innovative manner.

Certain case studies are elaborated in the Chap. 7 to highlight the potential of future research and innovations.

Chapter 7
Application of Sulphonation by Liquid SO$_3$ Dissolved in Liquid SO$_2$

7.1 Introduction

One of the first applications is suggested for the production of Sulphamic Acid (NH$_2$SO$_2$OH).

The present technique of producing Sulphamic Acid is by reacting oleum with Urea. The process is very cumbersome and hazardous.

The current process for manufacturing Sulphamic Acid employed by most of the producers requires the disposal of a large quantity of dilute Sulphuric Acid. The process to manufacture Sulphamic Acid was developed by the author in early sixties and it has gone under certain modifications in order to improve the economics, but it is still hazardous and requires disposal of waste acid. This is described in detail to highlight need for innovation. Please refer Bibliography section.

The new process that is developed as described in this chapter is not only to eliminate the above negative features, but also would reduce the CAPEX (CAPITAL EXPENDITURE) for the same capacity of Sulphamic Acid plant.

7.2 Properties of Sulphamic Acid

Physical Properties:

Property	Value
Mol. Wt	97.09
Mp, °C	205
Decomposition temperature, °C	209
Density at 25 °C, g/cm^3	2.126
Refractive indexes, 25 ± 3 °C	

(continued)

© The Author(s) 2016
N.G. Ashar, *Advances in Sulphonation Techniques*,
SpringerBriefs in Applied Sciences and Technology,
DOI 10.1007/978-3-319-22641-5_7

(continued)

Property	Value
α	1.533
β	1.563
γ	1.568
Solubility, wt%	
Aqueous	
At 0 °C	12.08
20 °C	17.57
40 °C	22.77
60 °C	27.06
80 °C	32.01
Non-aqueous, at 25 °C	
Formaldehyde HCHO	16.67
Methanol CH_3OH	4.12
Phenol C_6H_5OH	1.67
Acetone CH_3COCH_3	0.40
Ether $CH_5OC_2H_5$	0.01
71.8 % Sulphuric acid H_2SO_4	0.00

7.3 Process

A block diagram in Fig. 7.2 is attached herewith to describe the new process applied on liquid SO_2 and liquid SO_3. For comparison, a block diagram for the current process is also given in block diagram Fig. 7.1. The details of the current process are given below:

Urea and 23–25 % Oleum are fed at controlled rates to series of Reactors, which are cooled by chilled water/brine and cooling water. The reaction products are diluted by mixing with recycled mother liquor (available after separation of crystals of Sulphamic Acid). Temperature is controlled during mixing by chilled water/brine. Dilute acid stream (70 % Sulphuric Acid) is separated after the mixing operation and is sold to SSP/Alum manufacturers.

Following suggestions of Baumgartner, equimolar quantities of urea, sulphur trioxide, and sulphuric acid are reacted directly with Sulphamic acid:-

$$NH_2CONH_2 + SO_3 + H_2SO_4 \rightarrow 2\ NH_2SO_3H + CO_2$$

This is a strongly exothermic reaction. The process is carried out in two stages, based on the following reactions:

$$NH_2CONH_2 + SO_3 \rightarrow NH_2CONHSO_3H$$

$$NH_2CONHSO_3H + H_2SO_4 \rightarrow 2NH_2SO_3H + CO_2$$

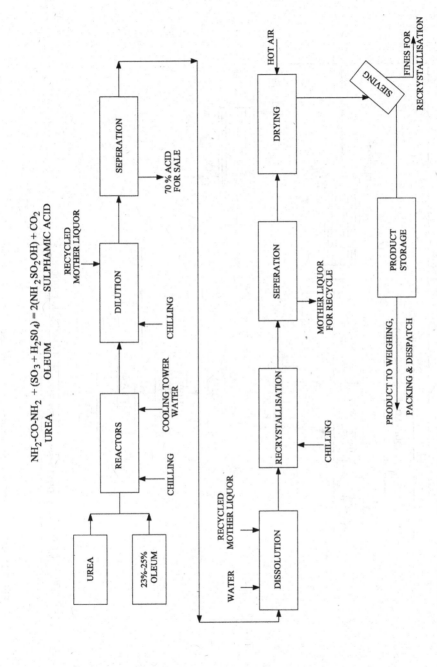

Fig. 7.1 Block diagram of sulphamic acid

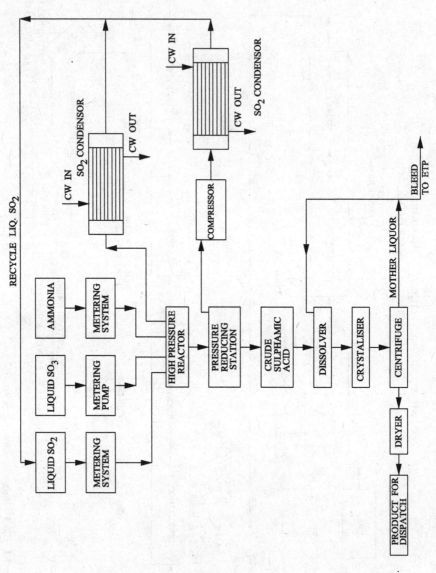

Fig. 7.2 Block diagram for manufacture of sulphamic acid without refrigeration

Sulphamic acid crystal (slurry) is dissolved in recycled mother liquor and makeup water is added as per need.

Recrystallation is carried out by further chilling and crystals of Sulphamic acid are separated out (which are dried by hot air), sieved and bagged. Fines obtained during sieving are redissolved and crystallized again.

Mother liquor obtained during separation of Sulphamic acid crystals is recycled as stated above.

The innovative process as described in Fig. 7.2 is as follows:-

Metered quantities of liquid SO_3, liquid SO_2 and liquid ammonia are fed to a high pressure agitated reactor with chilled water cooling.

The exothermic reaction of ammonia and sulphur trioxide forming Sulphamic acid by the equation

$$NH_3 + SO_3 \rightarrow NH_2SO_2(OH)$$

is removed by chilled water circulation as well as evaporation of liquid SO_2 acting as auto-refrigerant.

The evaporated sulphur dioxide is condensed by cooling tower water and recycled to the reactors via the liquid SO_2 storage and metering system.

An overflow of liquid SO_2 with crystals of sulphamic acid is fed to a pressure reducing reactor fitted with a jacket for heating,

The sulphur dioxide evaporated is compressed and cooled for liquefaction. This liquid SO_2 is recycled into the liquid SO_2 storage tank. The crystals of crude Sulphamic acid are fed to a dissolver with heating arrangement. Dissolver is fed with DM water for makeup and recycled mother liquor.

The solution of Sulphamic acid is fed to crystallizer section. The slurry of recrystallized Sulphamic Acid crystals are fed to a continuous centrifuge and flash drying system.

The product is then packed and sent for dispatch.

The main features of this process are

1. Absence of a chain of reactors fitted with chilling systems to prevent runaway reactions between oleum and urea.
2. Absence of requirement to produce 70 % sulphuric acid in order to crystallise crude sulphamic acid.
3. No stack emissions of carbon dioxide with traces of sulphur trioxide and sulphuric acid.
4. Minimum load of refrigeration system
5. Minimum utility consumption and load on effluent treatment plant.
6. Substantial reduction in capital cost.
7. Lower cost of raw materials
8. High efficiency due to reaction at sub zero temperatures.

7.4 Uses

It is estimated that world annual production of Sulphamic Acid is 60,000 tonnes. Most of this is used in cyclamate production (sweeteners). The use of Sulphamic acid in cleaning agents for carbonate and phosphate containing deposits, e.g. boiler scale, is based on its ability to form readily soluble salts and its relatively low corrosive effect on metals.

Sulphamic acid is widely used for cleaning machines and instruments in the paper, sugar, dairy and brewing industries, and for removing deposits in evaporation plants, heat exchangers, and cooling systems.

In some countries, a process involving the treatment of fatty or ethoxylated alcohols with Sulphamic acid to produce raw materials for waxes is used on an industrial scale.

Paper Pulp Bleaching:- Sulphamic acid additions to chlorination bleaching stages are effective in reducing pulp-strength degradation associated with high temperatures. Other benefits are noted when sulphamic acid is added to the hypo chloride bleaching stage, including reduction of pulp-strength losses as a result of high temperature or low pH; increased production by means of higher temperatures and lower pHs at the same pulp-strength level; savings in chemical costs, e.g. Lower consumption of buffer, caustic soda, and higher priced bleaching agents; improved efficiency through reducing effects of variation in temperature and pH.

Chlorine Vehicle and Stabilizer:- Sulphamic acid reacts with hypochlorous acid to produce N-chlorosulphamic acids, compounds in which the chlorine is still active but more stable than in hypochorite form. The commercial interest in this area is for chlorinated water systems in paper mills, i.e. for slimcides, cooling towers and similar applications.

Analytical and Laboratory Operations:- Sulphamic acid has been recommended as a reference standard in acidimetry. It can be purified by recrystallization to give a stable product that is 99.95 wt% pure. The reaction with nitrite as used in the sulphamic acid analytical method has also been adapted for determination of nitrites with the acid as the reagent. This reaction is used commercially in other systems for removal of nitrous acid impurities, e.g. in sulphuric and hydrochloric acid purification operations.

Sulphation and Sulphamation:- Sulphamic acid can be regarded as an ammonia-SO_3 complex and has been used thus commercially, always in anhydrous systems. Sulphation of mono, i.e. primary and seconday; polyhydric alcohols; unsaturated alcohols; phenols; and phenol ethylene oxide condensation products has been performed with sulphamic acid. The best-known application of sulphamic acid for sulphamation is the preparation of sodium cyclohexylsulphamate, which is a synthetic sweetener.

7.5 Conclusion

The process described in this chapter is only one of the applications. In Chap. 9, few more case studies are described for the sulphonation of organic chemicals using properties of liquid SO_2 and liquid SO_3.

Chapter 8
Impact on the Future Processes for the Manufacture of Chemicals

8.1 Introduction

The applications mentioned in Chap. 7 have far reaching implications on sulphonation processes to produce complex chemicals, which hitherto have been manufactured by costly processes, involving various time consuming steps and requiring disposal of hazardous chemicals. The subsequent Chap. 9 will illustrate the above by taking up case studies. It is required that considerable amount of research work to be carried out in order to supply similar techniques for other chemicals specially in the detergent, pharmaceutical, dyestuff and specialty chemicals for petroleum refinery.

In order to illustrate the impact on future processes we are pleased to give herewith the following comparison to produce valuable product widely used namely Para Toluene Sulphonic Acid (PTSA). The current process for manufacturing this chemical is sulphonation of toluene at high temperatures in a glass lined reactors using 98 % Sulphuric acid.

The new process involved would be at low temperature and would not require any Sulphuric Acid for the sulphonation of toluene to produce Para Toluene Sulphonic Acid (PTSA).

1

A brief process flow chart is attached in the Fig. 8.1.

8.2 Raw Materials Required

Raw Material Consumption/MT of the Product

		Cas no.	Qty (MT)
1	Toluene	[108-88-1]	0.48
2	Sulphuric Acid	[7664-93-9]	0.52

© The Author(s) 2016
N.G. Ashar, *Advances in Sulphonation Techniques*,
SpringerBriefs in Applied Sciences and Technology,
DOI 10.1007/978-3-319-22641-5_8

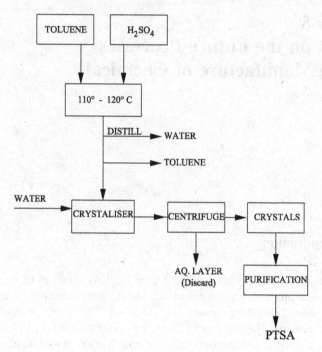

Fig. 8.1 Process currently employed to manufacture para toluene sulphonic acid

8.3 Major Areas in Which Experimental Work Should Be Directed

1. Sulphamic Acid
2. H-Acid
3. Napthalene Sulphonic Acid
4. Dry intermediates involving sulphonation
5. Speciality chemicals for petroleum refinery

8.4 Specifications (PTSA Monohydrate)

	Parameters	Limits
1	Appearance	White crystalline material
2	Assay	>98.5 %
3	Impurities	<0.5 % H2SO4
		~9.1 % water
4	m.p	103–106 °C

8.5 Commercial Details Current Manufacturers in India

(1) Hrishabh Life Sciences
(2) Nandadeep Chemicals Pvt.Ltd.,
(3) Navdeep Chemicals Pvt Ltd.,
(4) Nisha Chemicals
(5) Paresh Chemicals
(6) Reena Chem
(7) Sanjay Chemicals(India) Pvt Ltd.,
(8) Vertex Dyechem Corporation

8.6 Applications and End Use

- PTSA is comparable in strength to mineral acids such as Sulphuric Acid, but are especially suitable for organic reactions where an inorganic, mineral acid could cause charring, oxidation, or an unwanted chemical reaction.
- It is most useful as acid-catalyst reaction such as esterification/condensation/acetylation/polymerization/alkylation/hydrolysis/dehydration. As a hardening agent of furan resin which is applied in foundry Industry (Sand Casting).
- In Plasticizers industries mainly for DEP/DOP/DBP—As esterification catalyst. It is also used as curing agent for many resin system such as amino/phenolic/acrylic resins/cresols/epoxies/amino-Plastics and furniture lacquers.
- As an anti-stress additive for electroplating baths additive, plastics, coatings, dyes, pharmaceutical intermediates, hydrotrope, coupling agent and as a wetting agent.
- Also applicable for use in food packaging adhesives.
- As intermediate of dye chemistry
- In Butyl Acrylate manufacturing—As catalyst In Pharmaceutical industries is used for manufacturing of below product—As catalyst. CIPROFLOXACIN, Persantin, Naproxen DOXYCYCLIN SULPHAMETHAXAZOLE. And produce the intermediates of Amoxicllion, Cefadroxil.

8.7 Effluent Expected

2 MT (aqueous acidic).

Fig. 8.2 Neat's innovative
process to produce para
toluene sulphonic acid

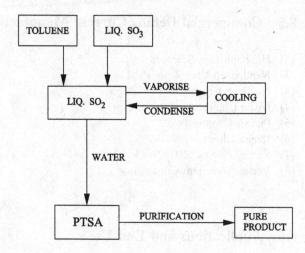

8.8   Alternate Process (Proposed)

The innovative process using liquid SO_2 and liquid SO_3 avoids elaborate chain of
glasslined reactors with thermic fluid heating and vacuum distillation batch systems
in Fig. 8.2.

8.9   Key Physical Properties for the New Process

		m.p (°C)	b.p (°C)
1	Toluene	−95	111
2	SO_3	16.9	45
3	SO_2	−72	−10

Advantages of the new process

1. Continous reaction hence small capital investment as well as power requirement.
2. Low temperature reaction hence better conversion and better quality.
3. Pharmaceutical grade can be produced by dissolving and recrystallization.
4. No effluent generation solid, liquid or gaseous.

Chapter 9
Case Studies and Its Commercial Application

9.1 Introduction

In this chapter, we propose to highlight case studies of two valuable chemicals which are being produced in bulk throughout the world

1. Para Toluene Sulphonic Acid widely used as plasticizers, additive in foundry chemicals, production of cresols, additive in detergents as sodium salts and other chemical reactions. These are outlined as below:-
2. Production of Raw material to manufacture saccharine namely otho toluene sulphonamine. The byproduct is para toluene sulphonic chloro amine—T for water purification.

At present the world production of Saccharine is in the range of 6000–8000 tonnes/year. The world requires minimum of 100,000 tonnes of Chloro Sulphonic Acid for sulphonation of toluene, a series of batch reactors requiring chilling unit for better conversion of toluene orther isomer. It will also produce 150,000 tonnes of hydro chloric acid (commercial grade 30 %) the innovative process will avoid the above drawbacks in the current manufacture and thereby produce saccharine at much cheaper price.

© The Author(s) 2016
N.G. Ashar, *Advances in Sulphonation Techniques*,
SpringerBriefs in Applied Sciences and Technology,
DOI 10.1007/978-3-319-22641-5_9

9.2 NEAT's Innovative "Cold Process"

As regards failures and plant stoppages from use of conventional Sulphuric acid, equipment as practiced today are elaborated in our earlier publication of springer title "A Practical insight in the Manufacture of Sulphuric Acid, Oleums & Sulphonating Agents."

As regards tabular comparison between NEAT's DCDA (3+2) process and conventional DCDA (3+1) process NEAT has already under implementation projects for 3 plants in India and one plant in Saudi Arabia to use this innovative process to produce liquid SO_3. The capacities envisaged are 52 TPD, 60 TPD, 60 TPD and 20 TPD.

9.3 Para Toluene Sulphonic Acid

Name	PTSA
Synonyms	Para Toluene sulphonic acid 4-Methylbenzene sulphonic acid
CAS No.	[104-15-4] anhydrous [6192-52-5] monohydrate
Molecular weight	172.2 (Anhydrous) 190.22 (Monohydrate)
Molecular formula	$C_7H_8SO_3$ (Anhydrous) $C_7H_{10}SO_4$ (Monohydrate)
Structural formula	

9.4 Synthesis (Outline)

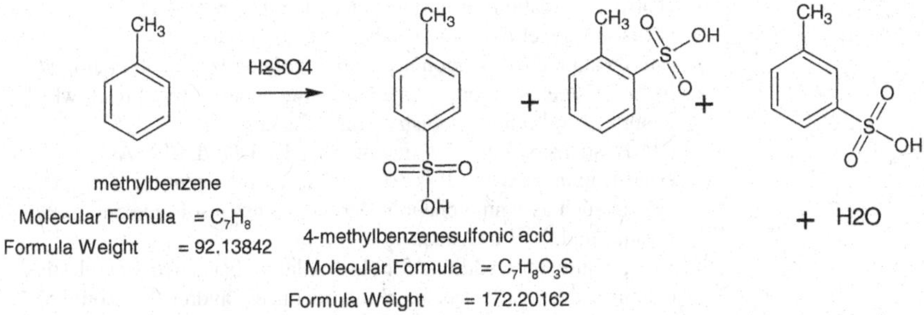

methylbenzene

Molecular Formula = C_7H_8

Formula Weight = 92.13842

4-methylbenzenesulfonic acid

Molecular Formula = $C_7H_8O_3S$

Formula Weight = 172.20162

+ H2O

9.5 Specifications (PTSA Monohydrate)

	Parameters	Limits
1	Appearance	White crystalline material
2	Assay	>98.5 %
3	Impurities	<0.5 % H_2SO_4 ~9.1 % water
4	m.p	103–106 °C

9.6 Commercial Details

a	Price	$1500–2000/MT
b	Manufacturers (India)	(1) Hrishabh Life Sciences
		(2) Nandadeep Chemicals Pvt.Ltd
		(3) Navdeep Chemicals Pvt Ltd.,
		(4) Nisha Chemicals
		(5) Paresh Chemicals
		(6) Reena Chem
		(7) Sanjay Chemicals(India) Pvt Ltd.,
		(8) Vertex Dyechem Corporation

(continued)

(continued)

c	Application/end use	• PTSA is comparable in strength to mineral acids such as Sulphuric Acid, but are especially suitable for organic reactions where an inorganic, mineral acid could cause charring, oxidation, or an unwanted chemical reaction • It is most useful as acid-catalyst reaction such as esterification/condensation/acetylation/polymerization/alkylation's/hydrolysis/dehydration. · As a hardening agent of furan resin which is applied in foundry Industry (Sand Casting) • In Plasticizers industries mainly for DEP/DOP/DBP—As esterification catalyst.· It is also used as curing agent for many resin system such as amino/phenolic/acrylic resins/cresols/epoxies/amino-Plastics and furniture lacquers • As an anti-stress additive for electroplating baths additive, plastics, coatings, dyes, pharmaceutical intermediates, hydrotrope, coupling agent and as a wetting agent • Also applicable for use in food packaging adhesives • As intermediate of dye chemistry • In Butyl Acrylate manufacturing—As catalyst. In Pharmaceutical industries is used for manufacturing of below product—As catalyst. CIPROFLOXACIN, Persantin, Naproxen DOXYCYCLIN SULPHAMETHAXAZOLE. And produce the intermediates of Amoxicillin, Cefadroxil
d	Effluent expected	∼2 MT(aqueous acidic)

9.7   Some Key Physical Parameters

		m.p (°C)	b.p (°C)
1	Toluene	−95	111
2	SO_3	16.9	45
3	SO_2	−72	−10

9.8   Important Inferences

1. Toluene is in liquid form in a wider range of temperatures −95 to 110 °C and will be useful as a solvent in the above reaction.
2. Sulphur dioxide will be in liquid form from −10 to −72 °C and sulphur trioxide will be in solid form under this condition.
3. Toluene and sulphur trioxide will be in liquid form around ambient temperature (i.e. around 25–30 °C)and may be easy to handle.

9.9 Techno Commercial Advantages of NEAT's Innovative Process to Manufacture PTS Acid

1. Continous plant instead of batch
2. Single reactor
3. Low capital and labour cost
4. DCS control possible
5. Higher efficiency
6. Low raw material and utility consumption per tonne of product
7. Better quality, pharma grade using DM water for crystallization
8. Large size of crystals

9.10 Case Study for Innovative Process Invented by NEAT to Carry Out Chloro Sulphonation of Toluene Without Refrigeration and CSA Plant as Raw Material to Manufacture Saccharine

9.10.1 Introduction

The world production of saccharine is estimated at approximately 60,000 tons per year. 10 % of this production is carried out by conventional process in India.

This process developed in the sixties and seventies is requiring high capacity chilling plant. Number of batch reactors, recovery of by product HCl gas producing large quantities of dilute Hydro chloric acid, low efficiency of conversion to ortho isomer and waste disposal in the effluent treatment plant.

9.11 Conventional Process

Description: Chloro Sulphonic Acid is taken into agitated reactors having jacket for cooling with chilled brine. A large capacity chilling plant is required to maintain sub-zero reaction temperatures to produce toluene sulphonyl chloride with more of ortho isomer compared to para.

PTS Chloride (Para Toluene Sulphonyl Chloride) is solid whereas ortho toluene sulphonyl chloride is liquid.

After the reaction is completed adding toluene gradually to maintain low temperature and remove exothermic heat of reaction OTS chloride is separated and PTS chloride is centrifuged and sent for drying.

For example, for production of 20 tonnes of saccharine per day one requires four reactors of 5000 litres each taking more than 18 h of reaction. For every tonne of chloro consumed one tonne of chemical grade (30 %) hydro chloric acid is produced.

9.12 A Brief Description of the Innovative Process Is to Affect the Drawbacks in the Conventional Process Is Indicated in Fig. 9.1. the Main Features of the Process Are as Under

A single reactor having liquid SO_2 under pressure (6–8 kg/cm^2) which is cooled to −10 °C or lower by chilled brine. This operation is for the initial start. The entire reaction being continuous the size of reactor can be as low as 1500–2000 L.

The high pressure reactor is fed with liquid SO_3, toluene and chloro sulphonic acid by metering pumps. The reaction takes place in two phases:

$$C_7H_8 + ClSO_2(OH) \rightarrow C_7H_7SO_2(OH) + HCl$$
$$C_7H_7SO_2(OH) + ClSO_2OH \rightarrow C_7H_7SO_2Cl + H_2O$$

The above reactions are highly exothermic and heat will be removed by evaporation of liquid Sulphur dioxide. The evaporated Sulphur dioxide is made up by flow meter from liquid SO_2 storage over flow is fed to a pressure vessel which evaporates excess sulphur dioxide to be condensed and recycled using cooling water at ambient temperatures from cooling towers. From the bottom of the reactor the reacted mass is sent to the evaporator. The reacted mass is sent to the crystallizer and centrifuge to give solid PTS chloride and liquid OTS chloride. This low pressure vessel is jacketed for providing heat to evaporated sulphur dioxide which is sent to catalytic converter to produce liquid SO_3 as a raw material producing Liquid sulphur trioxide.

9.13 Economics

Currenlty in India saccharine is sold at Rs. 700 (USD 110) per kg. The cost of the production is very high in view of the fact that one has to use twice the quantity of chloro sulphonic acid to the tune of 16 kg/kg of saccharine cost of which is equivalent to Rs. 150 (USD 2.25) per kg. In addition there is a high cost of utilities specially the chilling plant and operators manhours since the current process is batchwise.

In addition the requirement of higher capital cost as well as plant area and maintenance increase the cost of production.

The innovative process invented by NEAT can address the above drawbacks by having the following economic advantages (See Fig. 9.2):-

1. The process is continuous and hence smaller size single reactor
2. Absence of chilling plant due to auto refrigeration by evaporation of liquid SO_2
3. Lower sub-zero temperatures giving higher yield

Innovative Process for Chloro Sulphonation of Toluene Without
Refrigeration & CSA Plant as Raw Material to Manufacture Sachharin

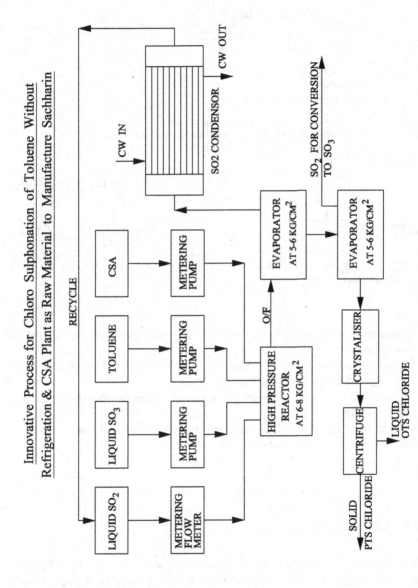

Fig. 9.1 Block diagram for innovative sulphonation

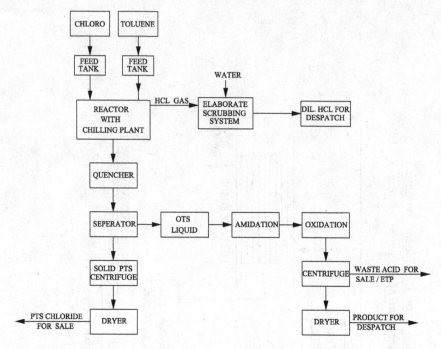

Fig. 9.2 Block diagram of manufacturing sachharin by conventional batch process of chlorosulphonation of toluene

4. Continous separation by pusher type centrifuge requiring lower capital cost and lower manhours
5. In situ conversion of hydrochloric gas evolved during reaction into chloro sulphonic acid required in the reaction
6. Recycle of liquid SO_2 by condensing at ambient temperatures using water from cooling towers
7. Production of liquid SO_3 from recovered sulhphur dioxide at low pressures for recycling in the high pressure continuous reactor using metering pumps.
8. The entire system can be controlled by Digital Cotnrol System (DCS)

9.14   Conclusion

In this chapter we have highlighted only two case studies. However, NEAT is in the process of developing similar applciations taking advantage of the properties of liquid SO_2 and liquid SO_3.

15 TPD liq SO_2 plant under construction by NEAT at M/s. Amal Rasayan Ltd at Ankleshwar, GIDC, Gujarat.

A closer look at a 180 TPD S.A. plant at Ankleshwar, Gujarat of M/s. Shree Sulphuric Acid. In the foreground one can see the Trombone Cooler for 25 % oleum.

A Closeup of high pressure reactor for 20 TPD Liquid SO_2 plant of M/s. Nath Industrial Chemicals Limited—VAPI, GIDC, India.

A Closeup of Liquid SO_2 Condensers at 14 M Level of M/s. Nath Industrial Chemicals Limited—VAPI, GIDC, India.

An overview of liquid SO_2 plant 20 TPD of M/s. Nath Industrial Chemicals Limited—VAPI, GIDC, India.

Duct from Economiser at Belagula plant of M/s. Bangur Fertiliser Chemicals in June 1992. In the foreground you can see the acid and cooler pipe with support rack and basins.

Liquid SO_2 plant installation being carried out for 15 TPD liquid SO_2 plant for M/s. Amal Rasayan Ltd in 2004–05.

Oleum tower cum tank for 65 and 25 % oleum a unique design by N G Ashar since 1965.

Chapter 10
Summary

This document is a path breaking narration for future technologies to carry out complex reactions of sulphonation. The book will be a guide to students of chemical engineering, technocrats and researcher's exploring new techniques to produce sulphur based chemicals using liquid sulphur trioxide and liquid sulphur dioxide.

This document has highlighted the importance of liquid sulphur dioxide and liquid sulphur trioxide for carrying out sulphonation reactions in an innovative way that would not only produce valuable products economically but also would be environment friendly.

The world production of over 150 million tonnes/year is affecting flora and fauna by acid rain of one million tonnes every year. This document indicates a 'Cold Process' which will produce Sulphuric Acid with Zero Emission of SO_2, one-half CAPEX, one-third area, less than half utility consumption. Also, co-generation of steam upto 1.4 tonnes/tonne of acid produced will be achieved against 1.15 tonnes/tonne SA.

The author strongly recommends that all existing and future sulphuric acid plants should adopt the 'Cold Process' for achieving lower capital, utility & maintenance costs, requiring one-third area, Zero Emission of sulphur dioxide and higher cogeneration of steam.

© The Author(s) 2016

N.G. Ashar, *Advances in Sulphonation Techniques*,
SpringerBriefs in Applied Sciences and Technology,
DOI 10.1007/978-3-319-22641-5_10

Appendices

1. Recent Trends in control of Sulphur dioxide emission for the manfacture of sulphuric acid in India specially with Co-generation of Power, N G Ashar–Abu Dhabi 1995
2. Liquid Sulphur Dioxide Without Compression or Refrigeration, N G Ashar–Calgary 1999
3. Comparative Study of Techno-Economic Evaluation of the Production of Liquid Sulphur Dioxide, N G Ashar–Berlin 2012
4. Indian Standards of liquid SO_2
5. Indian Standards of liquid SO_3
6. References

See Tables A.1 and A.2.

© The Author(s) 2016
N.G. Ashar, *Advances in Sulphonation Techniques*,
SpringerBriefs in Applied Sciences and Technology,
DOI 10.1007/978-3-319-22641-5

Table A.1 MB and EB for sulphuric acid plant [Alt 1-(65 TPD Sulphuric Acid + 12 TPD liq. SO$_3$) Alt 2-80 TPD]

Alt-1 (65 TPD Sulphuric acid + 12 TPD liq. SO$_3$)

St. No.	1	2	3	4	5	6	7	8
St. No. as per PFD	LS-1	LS-2	LS-4	LS-5	LS-6	LS-8	LS-9	LS-11
Location	Pump discharge APT-1	Pump discharge APT-1–PHE-1	PHE-1–IPAT	IPAT–APT-1	Pump discharge APT-1–PHE-2	PHE-2–D.T.	D.T.–APT-1	PHE-2 out–APT-2
Fluid	S. Acid	S. Acid	S. Acid	S. Acid	S. Acid	S. Acid	S. Acid	S. Acid
Concentration (%)	98.5	98.5	98.5	98.84	98.5	98.5	98.11	98.5
Flow rate (m^3/h)	121.419	80	80	80.7	41.419	40	40.379	1.419
Flow rate (kg/h)	222197.846	146400	146400	148488.88	75797.846	73200	73490.481	2597.846
Sp. Gravity	1.83	1.83	1.83	1.84	1.83	1.83	1.82	1.83
Temp (°C)	86.5	86.5	75	94.49	86.5	65	69.66	65
Heat Capacity (kcal/kg °C)	0.364	0.364	0.364	0.364	0.364	0.364	0.363	0.364
Heat flow (kcal/h)	4569.721 × 10^3	3010.862 × 10^3	2398.032 × 10^3	3485.681 × 10^3	1558.858 × 10^3	932.568 × 10^3	1058.011 × 10^3	33.096 × 10^3
Pressure (kg/cm^2)	3	3	1.8	By gravity	3	1.8	By gravity	2

St. No.	9	10	11	12	13	14	15	16
St. No.as per PFD	WS-2	LS-12	LS-13	LS-15	LS-16	LS-17	LS-19	WS-3
Location	Dilution Tank to APT-1	Pump Discharge APT-2	Pump Discharge APT-2 to PHE-3	PHE-3 to FAT	FAT to APT-2	PHE-3 to PHE-4	PHE-4 out to Storages	Dilution Tank to APT-2
Fluid	Water	S. Acid	S. Acid	S. Acid	S. Acid	S. Acid	S. Acid	Water
Concentration (%)	–	98.5	98.5	98.5	98.543	98.5	98.5	0.03226
Flow rate (m^3/h)	0.509	41.509	41.509	40	40.072	1.5095	1.5095	32.26
Flow rate (kg/h)	508.966	75962.506	75962.506	73200	73332.4	2762.506	2762.506	
Sp. Gravity	1	1.83	1.83	1.83	1.83	1.83	1.83	1
Temp (°C)	30	85.77	85.77	75	85.77	75	50	30
Heat Capacity (kcal/kg °C)	–	0.364	0.364	0.364	0.364	0.364	0.364	–
Heat flow (kcal/h)	–	1542.060 X 10^3	1542.060 X 10^3	1199.016 X 10^3	954.808 X 10^3	45.250 X 10^3	20.111 X 10^3	–
Pressure (kg/cm^2)	By gravity	3	3	1.8	By gravity	2	1.8	By gravity

Table A.1 (continued)

MB & EB—Sulphuric acid

Alt-2 (80 TPD Sulphuric Acid)

St. No.	1	2	3	4	5	6	7	8
St. No.as per PFD	LS-1	LS-2	LS-4	LS-5	LS-6	LS-8	LS-9	LS-11
Location	Pump discharge APT-1	Pump discharge APT-1–PHE-1	PHE-1–IPAT	IPAT–APT-1	Pump discharge APT-1–PHE-2	PHE-2–D.T.	D.T.–APT-1	PHE-2 out–APT-2
Fluid	S. Acid	S. Acid	S. Acid	S. Acid	S. Acid	S. Acid	S. Acid	S. Acid
Concentration (%)	98.5	98.5	98.5	98.92	98.5	98.5	98.11	98.5
Flow rate (m³/h)	121.759	80	80	80.972	41.759	40	40.379	1.759
Flow rate (kg/h)	222819.673	146400	146400	148988.88	76419.673	73200	73490.481	3219.673
Sp. Gravity	1.83	1.83	1.83	1.84	1.83	1.83	1.82	1.83
Temp (°C)	89.051	89.051	75	98.48	89.051	65	69.66	65
Heat Capacity (kcal/kg °C)	0.364	0.364	0.364	0.364	0.364	0.364	0.363	0.364
Heat flow (kcal/h)	4789.412×10^3	3146.804×10^3	2398.032×10^3	3713.804×10^3	1642.607×10^3	932.568×10^3	1058.011×10^3	41.019×10^3
Pressure (kg/cm²)	3	3	1.8	By gravity	3	1.8	By gravity	2

St. No.	9	10	11	12	13	14	15	16
St. No.as per PFD	WS-2	LS-12	LS-13	LS-15	LS-16	LS-17	LS-19	WS-3
Location	Dilution Tank to APT-1	Pump Discharge APT-2	Pump Discharge APT-2 to PHE-3	PHE-3 to FAT	FAT to APT-2	PHE-3 to PHE-4	PHE-4 out to Storages	Dilution Tank to APT-2
Fluid	Water	S. Acid	S. Acid	S. Acid	S. Acid	S. Acid	S. Acid	Water
Concentration (%)	–	98.5	98.5	98.5	98.543	98.5	98.5	–
Flow rate (m3/h)	0.6308	41.849	41.849	40	40.072	1.849	1.849	0.03226
Flow rate (kg/h)	630.793	76584.333	76584.333	73200	73332.4	3384.333	3384.333	32.26
Sp. Gravity	1	1.83	1.83	1.83	1.83	1.83	1.83	1
Temp (°C)	30	85.77	85.77	75	85.77	75	50	30
Heat Capacity (kcal/kg °C)	–	0.364	0.364	0.364	0.364	0.364	0.364	–
Heat flow (kcal/h)	–	1554.683×10^3	1554.683×10^3	1199.016×10^3	954.808×10^3	55.435×10^3	24.638×10^3	–
Pressure (kg/cm²)	By gravity	3	3	1.8	By gravity	2	1.8	By gravity

Table A.2 MB and EB for liquid SO₂ plant (15 TPD)

Sr. No.	Stream Nos.		1	2	3	4	5	6
	Description		DT-101 from storage tank of SAP	DT-101-T-103	DT-101-T-104	DT-103 from storage tank of SAP	DT-103-R-101	DT-102 from sulphur pit of SAP
	Component	Mol. wt						
1	Sulfuric acid 98.5 %	98	940 kg/month	470 kg/month	470 kg/month	–	–	–
2	Oleum 25 %	178	–	–	–	2600 kg/each startup	2600 kg/each startup	–
3	Liquid sulfur	32	–	–	–	–	–	4200 kg/48 h
4	Liquid sulfur-tri-oxide	80	–	–	–	–	–	–
5	Sulfur-di-oxide	64	–	–	–	–	–	–
6	Purge gas	–	–	–	–	–	–	–
7	Unreacted SO_3	80	–	–	–	–	–	–
8	Sulfur lumps	32	–	–	–	–	–	–
9	Acid mist	98	–	–	–	–	–	–
10	Strong acid	98	–	–	–	–	–	–
11	Spent acid	–	–	–	–	–	–	–
12	NaOH lye 10 %	40	–	–	–	–	–	–
13	Na_2SO_3 15 %	104	–	–	–	–	–	–
14	Moisture		–	–	–	–	–	–
15	Sulfur trioxide gas	80	–	–	–	–	–	–
16	Hot water							
17	Cooling water							
18	Steam							
	Total		940 kg/month	470 kg/month	470 kg/month	2600 kg/each startup	2600 kg/each startup	4200 kg/48 h
	Transfer time (h)		0.5	0.5	0.5	0.75	0.75	2
	Liquid flow (m³/h)		1.022	0.5110	0.5110	1.864	1.864	1.2
	Gas flow (m³/h)		–	–	–	–	–	–
	Temperature (°C)		35	35	35	35	35	125
	Pressure (kg/cm²g)		2.5	By gravity	By gravity	2.5	2.5	2.5
	Density (MT/m³)		1.84	1.84	1.84	1.86	1.86	1.75

(continued)

Table A.2 (continued)

Sr. No. / Stream Nos.	1	2	3	4	5	6
Description	DT-101 from storage tank of SAP	DT-101–T-103	DT-101–T-104	DT-103 from storage tank of SAP	DT-103–R-101	DT-102 from sulphur pit of SAP
Viscosity (Cp)	15.00	15.00	15.00	22.50	22.50	10.40
Sp. Heat (Kcal/Kg °C)	0.37	0.37	0.37	0.33	0.33	0.97
Phase	L	L	L	L	L	L
Enthalpy Kcal/h	3478.888	1739.444	1739.444	5720.616	5720.616	193515
Stream Nos.	7	8	9	10	11	12
Description	DT-102 when feeding to R-101 is off	DT-102–R-101	T-101 A(B) from V-101 A(B)	T-101 A(B) when feeding to R-101 is off	T-101 A(B)–R-101	R-101–T-106

Sr. No.	Component	Mol wt	1	2	3	4	5	6
1	Sulfuric acid 98.5 %	98	–	–	–	–	–	–
2	Oleum 25 %	178	–	–	–	–	–	–
3	Liquid sulfur	32	87.5 kg/h	87.5 kg/h	–	–	–	–
4	Liquid sulfur-tri-oxide	80	–	–	3364 kg/8 h	420.5 kg/h	420.5 kg/h	–
5	Sulfur-di-oxide	64	–	–	–	–	–	–
6	Purge gas	–	–	–	–	–	–	–
7	Unreacted SO$_3$	80	–	–	–	–	–	–
8	Sulfur lumps	32	–	–	–	–	–	–
9	Acid mist	98	–	–	–	–	–	–
10	Strong acid	98	–	–	–	–	–	–
11	Spent acid	–	–	–	–	–	–	2600 kg/each startup
12	NaOH lye 10 %	40	–	–	–	–	–	–
13	Na$_2$SO$_3$ 15 %	104	–	–	–	–	–	–

(continued)

Table A.2 (continued)

Sr. No.	Description	7	8	9	10	11	12
	Stream Nos.	DT-102 when feeding to R-101 is off	DT-102-R-101	T-101 A(B) from V-101 A(B)	T-101 A(B) when feeding to R-101 is off	T-101 A(B)-R-101	R-101-T-106
14	Moisture	–	–				–
15	Sulfur trioxide Gas	–	–	–	–	–	–
16	Hot water						
17	Cooling water						
18	Steam						
	Total	87.5 kg/h	87.5 kg/h	3364 kg/8 h	420.5 kg/h	420.5 kg/h	2600 kg/each startup
	Transfer time (h)	87.5 kg/h	87.5 kg/h	2	Continuous	Continuous	1
	Liquid flow (m^3/h)	0.05	0.05	0.876	0.219	0.219	1.398
	Gas flow (m^3/h)	–	–	–			–
	Temperature (°C)	125	125	35	35	35	40
	Pressure (kg/cm^{2}g)	1.5	7	By gravity	1.5	7	By gravity
	Density (MT/m^3)	1.75	1.75	1.92	1.92	1.92	1.86
	Viscosity (Cp)	10.40	10.40	1.50	1.50	1.50	18.00
	Sp. Heat (Kcal/Kg °C)	0.97	0.97	0.16	0.16	0.16	0.33
		L	L	L	L	L	L
	h	8063.125	8063.125	1345.536	336.4	336.4	8580.924
Sr. No.	Stream Nos.	13	14	15	16	17	18
	Description	R-101-PH-101	R-101-PLT-101 A	PLT-101 A From SAP	PLT-101 A-PH-101	PLT-101 A-PLT-101 B	PLT-101 B From SAP
	Component						

Sr. No.	Component	Mol. wt						
1	Sulfuric acid 98.5 %	98	–	–	–	–	–	–
2	Oleum 25 %	178	–	–	–	–	–	–
3	Liquid sulfur	32	–	–	–	–	–	–
4	Liquid sulfur-tri-oxide	80	–	–	–	–	–	–

(continued)

Table A.2 (continued)

Sr. No.	Description		13 R-101–PH-101	14 R-101–PLT-101 A	15 PLT-101 A From SAP	16 PLT-101 A–PH-101	17 PLT-101 A–PLT-101 B	18 PLT-101 B From SAP
	Stream Nos.							
5	Sulfur-di-oxide	64	–	505.624 kg/h	–	–	507.248 kg/h	–
6	Purge gas	–	0.250 kg/h	–	–	0.125 kg/h	0.699 kg/h	–
7	Unreacted SO_3	80	–	2.126 kg/h	–			
8	Sulfur lumps	32	–	–	210.24 kg/month		–	90 kg/month
9	Acid mist	98	–	–	–	–	–	–
10	Strong acid	98	–	–	–	–		–
11	Spent acid	–	–	–	–	–		–
12	NaOH lye 10 %	40	–	–	–	–		
13	Na_2SO_3 15 %	104	–	–	–	–		
14	Moisture		–	–	–	–		
15	Sulfur trioxide gas	80	–	–	–	–		
16	Hot water							
17	Cooling water							
18	Steam							
	Total		0.250 kg/h(2 kg/8 h)	507.75 kg/h	210.24 kg/month	0.125 kg/h (1 kg/8hrs)	507.947 kg/h	90 kg/month
	Transfer time (h)		5 s	Continuous	1	2.5 s	Continuous	1 h
	Liquid flow (m³/h)		–	–	Manually	–	–	Manually
	Gas flow (m³/h)		0.11	223.19	–	0.055	223.273	–
	Temperature (°C)		65	65	30	62	62	30
	Pressure (kg/cm²g)		6	6	–	5.8	5.8	–
	Density (MT/m³)		2.275 kg/m³	2.275 kg/m³	–	2.275 kg/m³	2.275 kg/m³	–
	Viscosity (Cp)		0.000849	0.0008490	–	0.000793	0.000793	–

(continued)

Table A.2 (continued)

Sr. No.	Stream Nos. Description		13	14	15	16	17	18
	Description		R-101-PH-101	R-101-PLT-101 A	PLT-101 A From SAP	PLT-101 A-PH-101	PLT-101 A– PLT-101 B	PLT-101 B From SAP
	Sp. Heat (Kcal/Kg °C)		0.15	0.15	–	0.15	0.15	–
	Phase		G	G	S	G	G	S
	Enthalpy Kcal/h		1.3125	2665.6875	–	0.6	2438.1456	–
	Stream Nos.		19	20	21	22	23	24
Sr. No.	Description		PLT-101 B-PH-101	PLT-101 B-T-102	T-102-T-103	T-103-PH-101	T-103-T-107	T-103-T-104
	Component	Mol. wt	–	–	–	–	–	–
1	Sulfuric acid 98.5 %	98	–	–	–	–	–	–
2	Oleum 25 %	178	–	–	–	–	–	–
3	Liquid sulfur	32	–	–	–	–	–	–
4	Liquid sulfur-tri-oxide	80	–	–	–	–	–	–
5	Sulfur-di-oxide	64	–	507.871 kg/h	507.871 kg/h	–	–	507.746 kg/h
6	Purge gas	–	0.125 kg/h	–	–	0.125 kg/h	–	–
7	Unreacted SO_3	80	–	0.046 kg/h	0.046 kg/h	–	0.144 kg/month	0.0458 kg kg/h
8	Sulfur lumps	32	–	–	–	–	–	–
9	Acid mist	98	–	–	–	–	–	0.002 kg/h
10	Strong acid	98	–	–	–	–	470 kg/month	–
11	Spent acid	–	–	–	–	–	–	–
12	NaOH lye 10 %	40	–	–	–	–	–	–
13	Na_2SO_3 15 %	104	–	–	–	–	–	–
14	Moisture	98	–	–	–	–	–	–
15	Sulfur trioxide gas	80	–	–	–	–	–	–
16	Hot water							
17	Cooling water							
18	Steam							

(continued)

Table A.2 (continued)

Sr. No.		19	20	21	22	23	24
Stream Nos.		19	20	21	22	23	24
Description		PLT-101 B-PH-101	PLT-101 B-T-102	T-102-T-103	T-103-PH-101	T-103-T-107	T-103-T-104
Total		0.125 kg/ht(1 kg/8 h)	507.917 kg/h	507.917 kg/h	0.125 kg/ht(1 kg/8 h)	470.144 kg/month	507.7938 kg/h
Transfer time (h)		2.5 s	Continuous	Continuous	2.5 s	0.5	Continuous
Liquid flow (m³/h)		–	–	–	–	0.511	–
Gas flow (m³/h)		0.055	223.26	NEAT	0.055	–	223.206
Temperature (°C)		62	62	60	56	40	56
Pressure (kg/cm²g)		5.6	5.6	5.5	5.5	By gravity	5.3
Density (MT/m³)		2.275 kg/m³	2.275 kg/m³	2.275 kg/m³	2.275 kg/m³	1.84	2.275 kg/m³
Viscosity (Cp)		0.000793	0.000793	0.000583	0.000583	14.00	0.000583
Sp. Heat (Kcal/Kg °C)		0.15	0.15	0.15	0.15	0.37	0.15
Phase		G	G	G	G	L	G
Enthalpy Kcal/h		0.6	2438.0016	1523.751	0.375	3478.888	1523.3814
Stream Nos.		25	26	27	28	29	30
Description		T-104-PH-101	T-104-T-107	T-104-ME-101 A	ME-101 A-T-103 or T-104	ME-101 A-ME-101 B	ME-101 B-T-103 or T-104
Sr. No.	Component	Mol. wt					
1	Sulfuric acid 98.5 %	98	–	–	–	–	–	–
2	Oleum 25 %	178	–	–	–	–	–	–
3	Liquid sulfur	32	–	–	–	–	–	–
4	Liquid sulfur-tri-oxide	80	–	–	–	–	–	–
5	Sulfur-di-oxide	64	–	–	507.621 kg/h	–	507.621 kg/h	–
6	Purge gas	–	0.125 kg/h	–	–	–	–	–
7	Unreacted SO₃	80	–	0.072 kg/month	0.0457 kg/h	–	0.0457 kg/h	–
8	Sulfur lumps	32	–	–	–	–	–	–
9	Acid mist	98	–	–	0.005 kg/h	0.003 kg/h	0.002 kg/h	0.007 kg/h
10	Strong acid	98	–	470 kg/month	–	–	–	–

(continued)

Table A.2 (continued)

Sr. No.	Stream Nos.		25	26	27	28	29	30
	Description		T-104–PH-101	T-104–T-107	T-104–ME-101 A	ME-101 A–T-103 or T-104	ME-101 A–ME-101 B	ME-101 B–T-103 or T-104
11	Spent acid	-	-	-	-	-	-	-
12	NaOH lye 10 %	40	-	-	-	-	-	-
13	Na$_2$SO$_3$ 15 %	104	-	-	-	-	-	-
14	Moisture		-	-	-	-	-	-
15	Sulfur trioxide gas	80	-	-	-	-	-	-
16	Hot water							
17	Cooling water							
18	Steam							
	Total		0.125 kg/ht(1 kg/8 h)	470.072 kg/month	507.6717	0.003 kg/h	507.6687 kg/h	0.007 kg/h
	Transfer time (h)		2.5 s	0.5	Continuous	-	Continuous	Continuous
	Liquid flow (m^3/h)		-	0.511	-	0.0016	-	0.00376
	Gas flow (m^3/h)		0.055	-	223.152	-	223.151	-
	Temperature (°C)		51	40	51	46	50	46
	Pressure (kg/cm^{2}g)		5.1	By gravity	5.1	5.1	4.9	4.9
	Density (MT/m^3)		2.275 kg/m^3	1.84	2.275 kg/m^3	1.86	2.275 kg/m^3	1.86
	Viscosity (Cp)		0.00055	14.00	0.00055	13.00	0.000534	13.00
	Sp. Heat (Kcal/Kg °C)		0.15	0.37	0.15	0.37	0.15	0.37
	Phase		G	L	G	L	G	L
	Enthalpy Kcal/h		0.3375	3478.888	1370.7136	0.01776	1294.555	0.03885
Sr. No.	Stream Nos.		31	32	33	34	35	36
	Description		ME-101 B–EX-102 A&B	EX-102 A&B–PH-101	EX-102 A&B–V-102 A(B)	V-102 A(B)–PH-101	V-102 A(B)–R-101	V-102 A(B)–EX-103 A & B
	Component	Mol. wt						
1	Sulfuric acid 98.5 %	98	-	-	-	-	-	-
2	Oleum 25 %	178	-	-	-	-	-	-
3	Liquid sulfur	32	-	-	-	-	-	-

(continued)

Table A.2 (continued)

Sr. No.	Stream Nos. Description		31 ME-101 B–EX-102 A&B	32 EX-102 A&B–PH-101	33 EX-102 A&B–V-102 A(B)	34 V-102 A(B)–PH-101	35 V-102 A(B)–R-101	36 V-102 A(B)–EX-103 A& B
4	Liquid sulfur-tri-oxide	80	–	–	–	–	–	–
5	Sulfur-di-oxide	64	507.621 kg/h	–	507.496 kg/h	–	–	507.371 kg/h
6	Purge gas	–	–	0.125 kg/h		0.125 kg/h	–	–
7	Unreacted SO_3	80	0.0457 kg/h	–	0.0457 kg/h	–	109.728 kg/months	0.0203 kg/h
8	Sulfur lumps	32	–	–	–	–	–	–
9	Acid mist	98	0.0013 kg/h	–	0.0013 kg/h	–	–	0.0013 kg/h
10	Strong acid	98	–	–	–	–	–	–
11	Spent acid	–	–	–	–	–	–	–
12	NaOH lye 10 %	40	–	–	–	–	–	–
13	Na_2SO_3 15 %	104	–	–	–	–	–	–
14	Moisture	–	–	–	–	–	–	–
15	Sulfur trioxide gas	80	–	–	–	–	–	–
16	Hot water							
17	Cooling water							
18	Steam							
	Total		507.668 kg/h	0.125 kg/h(1 kg/8 h)	507.543 kg/h	0.125 kg/h(1 kg/8 h)	109.728 kg/months	507.3926 kg/h
	Transfer time (h)		Continuous	2.5 s	Continuous	2.5 s	0.50	Continuous
	Liquid flow (m^3/h)		–	–	354.925	–	0.1143	–
	Gas flow (m^3/h)		223.151	0.055	–	0.055	–	222.03
	Temperature (°C)		47	47	36	45	36	45
	Pressure (kg/cm^2g)		4.7	4.7	4.3	4.3	4.3	5
	Density (MT/m^3)		2.275 kg/m^3	2.275 kg/m^3	1.43	2.275 kg/m^3	1.92	2.275 kg/m^3

(continued)

Table A.2 (continued)

Sr. No.		31	32	33	34	35	36
Stream Nos.							
Description		ME-101 B–EX-102 A&B	EX-102 A&B– PH-101	EX-102 A&B– V-102 A(B)	V-102 A(B)– PH-101	V-102 A(B)–R-101	V-102 A(B)– EX-103 A& B
Viscosity (Cp)		0.000501	0.000501	0.34 @ –10 °C and atm pressure	0.000501	1.50	0.000501
Sp. Heat (Kcal/Kg °C)		0.15	0.15	0.15	0.15	0.16	0.15
Phase		G	G	L	G	L	G
Enthalpy Kcal/h		1142.253	0.28125	456.7887	0.28125	210.678	1141.633
Stream Nos.		37	38	39	40	41	42
Description		EX-103 A&B– PH-101	EX-103 A&B– V-103 A(B)	PH-101–4th bed of cnvtr of SAP	PH-101–T-106	T-106–Phosphate Frtlzr plnt for sale	SC-101 from T-101 A(B), V-101A(B), EX-101 & T-106
Sr. No.	Component	Mol. wt					
1	Sulfuric acid 98.5 %	98	–	–	–	–	–	–
2	Oleum 25 %	178	–	–	–	–	–	–
3	Liquid sulfur	32	–	–	–	–	–	–
4	Liquid sulfur-tri-oxide	80				Traces		
5	Sulfur-di-oxide	64		507.246 kg/h				
6	Purge gas	–	0.125 kg/h		1.125 kg/h			
7	Unreacted SO₃	80		0.0203 kg/h				6 kg/ day
8	Sulfur lumps	32						
9	Acid mist	98	–	0.0013 kg/h				–
10	Strong acid	98						
11	Spent acid	–						
12	NaOH lye 10 %	40					2600 kg/each startup	
13	Na₂SO₃ 15 %	104						
14	Moisture							
15	Sulfur trioxide gas	80						

(continued)

Table A.2 (continued)

Sr. No.		37	38	39	40	41	42
Stream Nos.		37	38	39	40	41	42
Description		EX-103 A&B–PH-101	EX-103 A&B–V-103 A(B)	PH-101–4th bed of cnvtr of SAP	PH-101–T-106	T-106-Phosphate Frtlzr plnt for sale	SC-101 from T-101 A(B), V-101A(B), EX-101 & T-106
16	Hot water						
17	Cooling water						
18	Steam						
	Total	0.125 kg/h(1 kg/8 h)	507.2676 kg/h	1.125 kg/h(9 kg/8 h)	Traces	2600 kg/each startup	6 kg/ day
	Transfer time (h)	2.5 s	Continuous	0.5	10 s/7 days	2	6 h
	Liquid flow (m³/h)	–	354.7325	–	–	0.699	–
	Gas flow (m³/h)	0.055	–	0.4945	–	–	–
	Temperature (°C)	45	36	55	40	40	49
	Pressure (kg/cm²g)	5	4.3	3	5	2.5	ATM.
	Density (MT/m³)	2.275 kg/m³	1.43	2.275 kg/m³	1.920	1.86	2.84 kg/m³
	Viscosity (Cp)	0.000501	0.34 @ -10 °C and atm pressure	0.000669	1.20	22.00	0.90
	Sp. Heat (Kcal/Kg °C)	0.15	0.15	0.15	0.330	0.3	0.16
	Phase	G	L	G	L	L	G
	Enthalpy Kcal/h	0.28125	456.541	4.219	–	3900.000	3.04
	Stream Nos.	43	44	45	46	47	48
Sr. No.	Description	SC-101 From DT-101	SC-101–APT-2 of SAP	SC-102 from caustic 10 % NaOH lye storage	SC-102 from PH-101	SC-102–Storage of SAP SO2 scrubber	T-107–APT-2 of SAP

Component	Mol. wt						
1 Sulfuric acid 98.5 %	98	5778 kg/month	5818.5 kg/month	–	–	–	–
2 Oleum 25 %	178	–	–	–	–	–	–
3 Liquid sulfur	32	–	–	–	–	–	–
4 Liquid sulfur-tri-oxide	80	–	–	–	–	–	–

(continued)

Table A.2 (continued)

Sr. No.	Description		43	44	45	46	47	48
	Stream Nos.		SC-101 From DT-101	SC-101–APT-2 of SAP	SC-102 from caustic 10 % NaOH lye storage	SC-102 from PH-101	SC-102–Storage of SAP SO2 scrubber	T-107–APT-2 of SAP
5	Sulfur-di-oxide	64	–	–	–	–	–	–
6	Purge gas		–	–	–	1.125 kg/h (11.25 kg/10 h)	–	–
7	Unreacted SO_3	80	–	180 kg/month	–	–	–	–
8	Sulfur lumps	32	–	–	–	–	–	–
9	Acid mist	98	–	–	–	–	–	–
10	Strong acid	98	–	–	–	–	–	940.216 kg/month
11	Spent acid	–	–	–	–	–	–	–
12	NaOH lye 10 %	40	–	–	2486.644 kg/10 h SAP stoppage	–	–	–
13	Na_2SO_3 15 %	104	–	–	–	–	2497.894 kg/10 h	–
14	Moisture		40.5 kg/month	–	–	–	–	–
15	Sulfur trioxide gas	80	–	–	–	–	–	–
16	Hot water							
17	Cooling water							
18	Steam							
	Total		5818.5 kg/month	5998.5 kg/month	2486.644 kg/10 h SAP stoppage	1.125 kg/h (11.25 kg/10 h)	2497.894 kg/10 h	940.216 kg/month
	Transfer time (h)		1	1	1	1h	1	4
	Liquid flow (m³/h)		3.16	3.26	2.24	–	NEAT	0.1277
	Gas flow (m³/h)		–	–	–	0.4945	–	–
	Temperature (°C)		40	40	40	45	40	40
	Pressure (kg/cm²g)		2	2	1.5	2	1.5	2
	Density (MT/m³)		1.84	1.84	1.112	2.275 kg/m³	NEAT	1.84
	Viscosity (Cp)		14.00	14.00	1.21	0.000669	NEAT	14.00
	Sp. Heat (Kcal/Kg °C)		0.37	0.37	0.911	0.15	NEAT	0.37

(continued)

Table A.2 (continued)

Sr. No.		43	44	45	46	47	48
Description		SC-101 From DT-101	SC-101–APT-2 of SAP	SC-102 from caustic 10 % NaOH lye storage	SC-102 from PH-101	SC-102–Storage of SAP SO2 scrubber	T-107–APT-2 of SAP
Phase		L	L	L	G	L	L
Enthalpy Kcal/h		21517.35	22194.45	22653.327	4.21875	NEAT	869.7
Stream Nos.		49	50	51	52	53	54
Description		EX-101 from SAP	EX-101–V-101A/B	Hot Water from T-105–Jacket of R-101	Hot Water from Jacket of R-101–T-105	Hot Water from T-105-Jacket of T-102	Hot Water from Jacket of T-102–T-105
Sr. No.	Component	Mol. wt					
1	Sulfuric acid 98.5 %	98	–	–	–	–	–	–
2	Oleum 25 %	178	–	–	–	–	–	–
3	Liquid sulfur	32	–	–	–	–	–	–
4	Liquid sulfur-tri-oxide	80	–	–	–	–	–	–
5	Sulfur-di-oxide	64	–	–	–	–	–	–
6	Purge gas	–	–	–	–	–	–	–
7	Unreacted SO_3	80	–	–	–	–	–	–
8	Sulfur lumps	32	–	–	–	–	–	–
9	Acid mist	98	–	–	–	–	–	–
10	Strong acid	98	–	–	–	–	–	–
11	Spent acid	–	–	–	–	–	–	–
12	NaOH lye 10 %	40	–	–	–	–	–	–
13	Na_2SO_3 15 %	104	–	–	–	–	–	–
14	Moisture	–	–	–	–	–	–	–
15	Sulfur trioxide gas	80	500 kg/h	500 kg/h				
16	Hot water				50 m³/h	50 m³/h	16.667 m³/h	16.667 m³/h
17	Cooling water				–	–	–	–
18	Steam				–	–	–	–
	Total		500 kg/h	500 kg/h	50 m³/h	50 m³/h	16.667 m³/h	16.667 m³/h

(continued)

Table A.2 (continued)

Sr. No.	Stream Nos.	49	50	51	52	53	54
	Description	EX-101 from SAP	EX-101-V-101A/B	Hot Water from T-105-Jacket of R-101	Hot Water from Jacket of R-101-T-105	Hot Water from T-105-Jacket of T-102	Hot Water from Jacket of T-102-T-105
	Transfer time (h)	Continuous	Continuous	1 h	1 h	Continuous	Continuous
	Liquid flow (m³/h)	–	–	50	50	16.667 m³/h	16.667
	Gas flow (m³/h)	176.056	176.056	–	–	–	–
	Temperature (°C)	90	44	80	79.5	80	79.998
	Pressure (kg/cm²g)	0.5	By Gravity	2.5	2.2	2.5	2.3
	Density (MT/m³)	2.84 kg/m³	2.84 kg/m³	1	1	1	1
	Viscosity (Cp)	0.43	1.50	0.28	0.28	0.28	0.28
	Sp. Heat (Kcal/Kg °C)	0.16	0.16	1	1	1	1
	Phase	G	L	Liquid	Liquid	Liquid	Liquid
	Enthalpy Kcal/h	4800	1120	2500000	2475000	833350	833316.666
Sr. No.	Stream Nos.	55	56	57	58	59	60
	Description	Hot Water from T-105-Jacket of T-103	Hot Water from Jacket of T-103-T-105	Hot Water from T-105-Jacket of T-104	Hot Water from Jacket of T-104-T-105	Steam from Main header-T-105	Steam from Main header-T-108

	Component	Mol. wt	49	50	51	52	53	54
1	Sulfuric acid 98.5 %	98	–	–	–	–	–	–
2	Oleum 25 %	178	–	–	–	–	–	–
3	Liquid sulfur	32	–	–	–	–	–	–
4	Liquid sulfur-tri-oxide	80	–	–	–	–	–	–
5	Sulfur-di-oxide	64	–	–	–	–	–	–
6	Purge gas	–	–	–	–	–	–	–
7	Unreacted SO₃	80	–	–	–	–	–	–
8	Sulfur lumps	32	–	–	–	–	–	–
9	Acid mist	98	–	–	–	–	–	–
10	Strong acid	98	–	–	–	–	–	–

(continued)

Appendices

Table A.2 (continued)

Stream Nos.		55	56	57	58	59	60
Description		Hot Water from T-105–Jacket of T-103	Hot Water from Jacket of T-103–T-105	Hot Water from T-105–Jacket of T-104	Hot Water from Jacket of T-104–T-105	Steam from Main header–T-105	Steam from Main header–T-108
Sr. No.							
11	Spent acid	–	–	–	–	–	–
12	NaOH lye 10 %	–	–	–	–	–	–
13	Na$_2$SO$_3$ 15 %	–	–	–	–	–	–
14	Moisture	–	–	–	–	–	–
15	Sulfur trioxide gas	–	–	–	–	–	–
16	Hot water	16.667 m^3/h	16.667 m^3/h	16.667 m^3/h	16.667 m^3/h	–	–
17	Cooling water	–	–	–	–	–	–
18	Steam	–	–	–	–	36.419 kg/h	116.104 kg/h
	Total	16.667 m^3/h	16.667 m^3/h	16.667 m^3/h	16.667 m^3/h	36.419 kg/h	116.104 kg/h
	Transfer time (h)	Continuous	Continuous	Continuous	Continuous	Continuous	Continuous
	Liquid flow (m^3/h)	16.667 m^3/h	16.667	16.667 m^3/h	16.667	–	–
	Gas flow (m^3/h)	–	–	–	–	26.655	84.976
	Temperature (°C)	80	79.998	80	79.998	126.79	126.79
	Pressure (kg/cm^2 g)	2.5	2.3	2.5	2.3	1.5 (g)	1.5 (g)
	Density (MT/m^3)	1	1	1	1	0.7319 m^3/kg	0.7319 m^3/kg
	Viscosity (Cp)	0.28	0.28	0.28	0.28	0.14	0.14
	Sp. Heat (Kcal/Kg °C)	1	1	1	1	1	1
	Phase	Liquid	Liquid	Liquid	Liquid	Gas	Gas
	Enthalpy Kcal/h	833350	833316.666	833350	833316.666	19648.281	62637.666
Sr. No.	Stream Nos.	61	62	63	64	65	66
	Description	Hot Water from T-108–Jacket of V-102 A/B	Hot Water from Jacket of V-102 A/B–T-108	Cooling water from header–Jacket of R-101	Cooling water from Jacket of R-101–main Header (Return Line)	Cooling water from header–EX-101	Cooling water from EX-101–main Header (Return Line)
	Component (Mol. wt)						
1	Sulfuric acid 98.5 % (98)	–	–	–	–	–	–

(continued)

Table A.2 (continued)

Sr. No.	Description		61	62	63	64	65	66
	Stream Nos.		Hot Water from T-108-Jacket of V-102 A/B	Hot Water from Jacket of V-102 A/B-T-108	Cooling water from header-Jacket of R-101	Cooling water from Jacket of R-101-main Header (Return Line)	Cooling water from header-EX-101	Cooling water from EX-101-main Header (Return Line)
2	Oleum 25 %	178	-	-	-	-	-	-
3	Liquid sulfur	32	-	-	-	-	-	-
4	Liquid sulfur-tri-oxide	80	-	-	-	-	-	-
5	Sulfur-di-oxide	64	-	-	-	-	-	-
6	Purge gas	-	-	-	-	-	-	-
7	Unreacted SO_3	80	-	-	-	-	-	-
8	Sulfur lumps	32	-	-	-	-	-	-
9	Acid mist	98	-	-	-	-	-	-
10	Strong acid	98	-	-	-	-	-	-
11	Spent acid	-	-	-	-	-	-	-
12	NaOH lye 10 %	40	-	-	-	-	-	-
13	Na_2SO_3 15 %	104	-	-	-	-	-	-
14	Moisture		-	-	-	-	-	-
15	Sulfur trioxide gas	80	-	-	-	-	-	-
16	Hot water		50 m^3/h	50 m^3/h	15 m^3/h	15 m^3/h	20 m^3/h	20 m^3/h
17	Cooling water		-	-	-	-	-	-
18	Steam		-	-	-	-	-	-
	Total		50 m^3/h	50 m^3/h	15 m^3/h	15 m^3/h	20 m^3/h	20 m^3/h
	Transfer time (h)		Contonuous	Contonuous	Continuous	Continuous	Continuous	Continuous
	Liquid flow (m^3/h)		50	50	15	15	20	20
	Gas flow (m^3/h)		-	-	-	-	-	-
	Temperature (°C)		70	68.74	34	38	34	38
	Pressure (kg/cm^2g)		2.5	2.2	2.5	2.2	By gravity	By gravity
	Density (MT/m^3)		1	1	1	1	1	1

(continued)

Table A.2 (continued)

Sr. No.			61	62	63	64	65	66
	Stream Nos.		61	62	63	64	65	66
	Description		Hot Water from T-108-Jacket of V-102 A/B	Hot Water from Jacket of V-102 A/B-T-108	Cooling water from header-Jacket of R-101	Cooling water from Jacket of R-101-main Header (Return Line)	Cooling water from header-EX-101	Cooling water from EX-101-main Header (Return Line)
	Viscosity (Cp)		0.32	0.32	0.51	0.53	0.51	0.53
	Sp. Heat (Kcal/Kg °C)		1	1	1	1	1	1
	Phase		Liquid	Liquid	Liquid	Liquid	Liquid	Liquid
	Enthalpy Kcal/h		2000000	1937000	60000	120000	80000	160000
	Stream Nos.		67	68	69	70	71	72
Sr. No.	Description		Cooling water from header-EX-102 A & B each	cooling water from EX-102 A&B each-main Header (Return Line)	Cooling water from header-EX-103 A&B each	cooling water from EX-103 A&B each-main Header (Return Line)	Steam from header-jacket of DT-102	Condensate from jacket of DT-102-Recovery system
	Component	Mol. wt						
1	Sulfuric acid 98.5 %	98	–	–	–	–	–	–
2	Oleum 25 %	178	–	–	–	–	–	–
3	Liquid sulfur	32	–	–	–	–	–	–
4	Liquid sulfur-tri-oxide	80	–	–	–	–	–	–
5	Sulfur-di-oxide	64	–	–	–	–	–	–
6	Purge gas	–	–	–	–	–	–	–
7	Unreacted SO_3	80	–	–	–	–	–	–
8	Sulfur lumps	32	–	–	–	–	–	–
9	Acid mist	98	–	–	–	–	–	–
10	Strong acid	98	–	–	–	–	–	–
11	Spent acid	–	–	–	–	–	–	–
12	NaOH lye 10 %	40	–	–	–	–	–	–
13	Na_2SO_3 15 %	104	–	–	–	–	–	–

(continued)

Table A.2 (continued)

Sr. No.	Description		67	68	69	70	71	72
	Stream Nos.		Cooling water from header–EX-102 A & B each	cooling water from EX-102 A&B each–main Header (Return Line)	Cooling water from header–EX-103 A&B each	cooling water from EX-103 A&B each–main Header (Return Line)	Steam from header–jacket of DT-102	Condensate from jacket of DT-102–Recovery system
14		Moisture	–	–	–	–	–	–
15		Sulfur trioxide gas	80	–	–	–	–	–
16		Hot water	25 m³/h	25 m³/h	25 m³/h	25 m³/h	–	–
17		Cooling water	–	–	–	–	–	–
18		Steam					22.107 kg/h	22.107 kg/h
		Total	25 m³/h	25 m³/h	25 m³/h	25 m³/h	22.107 kg/h	22.107 kg/h
		Transfer time (h)	Continuous	Continuous	Continuous	Continuous	Contonuous	Contonuous
		Liquid flow (m³/h)	25	25	25	25	22.107 kg/h	22.107 kg/h
		Gas flow (m³/h)					11.801	
		Temperature (°C)	34	35	34	35	138.19	112.73
		Pressure (kg/cm²g)	2.5	2.3	2.5	2.3	2.5 (g)	0.6 (g)
		Density (MT/m³)	1	1	1	1	0.5338 m³/kg	1
		Viscosity (Cp)	0.51	0.51	0.51	0.51	0.10	0.15
		Sp. Heat (Kcal/Kg °C)	1	1	1	1	1	1
		Phase	Liquid	Liquid	Liquid	Liquid	Gas	Liquid
		Enthalpy Kcal/h	100000	125000	100000	125000	14422.6068	2497.2067

Notes

Steam requirement to T-105 to heat T-102,103 & 104 is only during winter season

Steam requirement for T-105 to heat Oleum in R-101 is only during each start-up

Peak Load of steam for T-105 to raise temperature of T-105, T-102,103 & 104 during winter for 1st hour of start-up in 1 h = 183481.149 Kcal/h, 318.807 kg/h @ 1.5 kg/cm2 (g) Pressure

Peak Load of steam for T-105 to raise temperature of T-105, V-102 A or B for 1st hour of start-up = 160657.1412 Kcal/h, 279.444 kg/h during winter @ 1.5 kg/cm2 (g) Pressure

Peak Load of steam for T-108 to raise temperature of T-108, V-102 A or B for 1st hour of start-up = 45931.838 Kcal/h, 78.166 kg/h @ 1.5 kg/cm2 (g) Pressure

Peak Load of steam to DT-102 to raise temperature of DT-102 for 1st hour of start-up = 17713.559 Kcal/h, 29.505 kg/h @ 2.5 kg/cm2 (g) Pressure

Bibliography

A practical insight in the manufacture of Sulphuric Acid, Oleums and Sulphonating agents—Published by the Author-2012

Duecker WW, West JR (1959) The manufacture of sulfuric acid. Robert E Krieger Publishing Co., Inc., Huntington, p 136

DKL Engg-Canada–material available on Internet

Gas Purification, 3rd edn. Arthur Kohl and Fred Riesenfeld, Gulf Publishing Co

Herman de Groot W (1991) Sulphonation technology in the detergent industry

Information available from Wikipedia

Monsanto Enviro Chem–for Dyna wave scrubbers

Own project work and plant visits, 1961–2012 by Author to SA plants in India, USA, Japan, Kenya, Thailand, Poland, USSR, Indonesia, etc.

Private communications with plant managers, engineers and owners, 2010–2012

Private communications and discussions with Dr. Matros on 'Unsteady state conversion' in USSR and USA in 1980's and 90's

Private communication with British Sulphur in 2008 on latest status of Sulphuric acid industry and sources of Sulphur

Perry's Chemical Engineer's handbook, 50th edn. McGraw Hill Publishing Co, pp 3–154

Sander U, Rothe U, Kola R (1984) Sulphur-sulphur dioxide and sulphuric acid. British Sulphur Corp Ltd and Verlag Chemie International, Inc, London

Sulphur, July–Aug 2004 issue no. 293 p 28

The chemical process industries, 2nd edn. McGraw Hill, p 149

Growth of Chlorosulphonic Acid Industry in India—CEW, November 1992. This publication being out of print Author has to represent the Graphs & Tables regarding properties of SO_2 & SO_3 & H_2SO_4. Contribution of Author is acknowledged with Thanks

© The Author(s) 2016

N.G. Ashar, *Advances in Sulphonation Techniques,*
SpringerBriefs in Applied Sciences and Technology,
DOI 10.1007/978-3-319-22641-5

Printed in the United States
By Bookmasters